RELATIVITY+

THE THEORY OF EVERYTHING

by

John Duffield

From basic concepts to Einstein's geometrical dream

Published in Great Britain in 2009 by Corella Limited
Registered office: Jonsen House, 43 Commercial Road, Poole, BH14 0HU
Website: www.corella-limited.com

First edition

A CIP record for this book is available from the British Library

ISBN: 978-0-9560978-0-4

Printed and bound in Great Britain by Caric Press Limited

INTRODUCTION 1

A BRIEF HISTORY OF RELATIVITY 10

MONEY EXPLAINED 16

THE PSYCHOLOGY OF BELIEF 20

ENERGY EXPLAINED 27

MASS EXPLAINED 36

GOING NOWHERE FAST 44

CHARGE EXPLAINED 47

REFERENCE FRAMES 57

TIME EXPLAINED 63

GRAVITY EXPLAINED 75

A NOTE ON NEWTON 88

SPACE EXPLAINED 89

EINSTEIN'S HEROES 98

PARTICLES EXPLAINED 102

QUANTUM MECHANICS 111

MORE PARTICLES 116

THE FINE STRUCTURE CONSTANT 122

THE STANDARD MODEL 125

THE UNIVERSE 131

INFLATION COMES FOR FREE 140

BLACK HOLES 143

THE BIG BANG 154

RELATIVITY+ 160

THE SAME ELEPHANT 165

REFERENCES 182

FIGURES 195

To the unsung heroes of science

INTRODUCTION

I've always held Einstein in the highest regard. I admire his ability to think outside the box, I empathise with his curiosity, and I share his desire to understand the world in terms we can grasp. And I so love the romance of the lowly patent clerk who applied his fresh young mind to the world's great mysteries. There he was in the dying days of horse-drawn transport, pondering his clocks and dreaming of space and time. What he came up with was just so profound. It feels like it was ahead of its time, and I suppose it was. It took a long time to catch on, a long time to gain acceptance, and a long time to catch the public eye. But when it did, the public were thrilled, as I am thrilled.

Figure 1 – Einstein at the patent office in Bern 1905

You find out more when you read the history. You learn that he didn't do it all, and it wasn't all right. You learn that many fine men and women are obscured by the shadow of his celebrity, and that things aren't quite as they seem. You learn that there have been shifts in interpretation that he wouldn't agree with, and that people say he said things that he didn't. You also learn of things forgotten, because he and others were ahead of their time, and the world wasn't ready. Because it's an imperfect world, because people aren't perfect. That's what you learn, that's how it is, that's how we are.

I ponder at what might have been. Perhaps Einstein or his successor would have achieved the dream of a unified field theory[1]. Perhaps fifty years ago somebody would have explained the postulates of special relativity, and told us why general relativity works the way it does. Perhaps this theory would have given us an intuitive understanding of electromagnetism, gravity and the other forces. Perhaps it would have explained wave/particle duality, quantum mechanics, and all the particles of the standard model. Perhaps it would have even explained black holes, dark energy, and the expansion of the universe. I like to think that the end product would have dispelled so much mystery that we could not have failed to have grasped how the universe works. And I like to think that as a result, the world would be a better place.

But nobody lives forever. The sands of time ran out for Albert Einstein, and nobody picked up the torch. The light of understanding slipped from our grasp, then spluttered and died. Ashes to ashes, dust to dust, life goes on. The world turns, days go by, years stretch into decades, time passes. And here I am. I've found his dream. I share his vision. I know where he was going in his twilight years, when he was sidelined, working alone. He wasn't just on his own, he was out on his own. But he never got it out. He got close. The signs are there when you know how to look. You see the resonance when you read what he said, not what people say he meant. You see the vision, you see the light. I've seen it, and I have to share it. It's beautiful, it's elegant, because less is more, because it's a tale of something and nothing. Once you share his vision, you see how simple it really is, and you wonder why you didn't see it before. I'll show you how to see it. But first I need to tell you how I got where I am. The story starts in 2005.

One fine day I was going about my business when I heard that the Maths tower at Manchester University was to be demolished. I was shocked. To me it was an inspirational icon of learning, the thing that put the campus on the skyline, on the map, and on the picture postcards. I used to sit on the seventeenth floor in a class, learning about gates and induction from the likes of Tom Kilburn[2], godfather of Computer Science. I listened intently with the cityscape before me and Jodrell Bank behind. I learned a lot in that Maths tower. It shaped my life, it was my beacon. And now it's gone.

Figure 2 – Manchester University Maths tower, now demolished

Yes, I know buildings don't last forever. But the building that replaced it is a grey pillbox lecture theatre for commercial seminars, part of a social "sciences" development, which in turn is part of a new vision for a 21st century campus. In this vision, the new mathematics building isn't the Maths Building. It's shared with other schools. It's the Astronomy, Mathematics, Physics and Photon Sciences building, formerly known as AMPPS, now named in honour of Alan Turing. And it's on the edge of campus. Yes I'm sure it's more modern, and more practical. But it's out on a limb. It just isn't so important any more. And moreover, UMIST is no more. The *University of Manchester Institute of Science and Technology* no longer exists. It's been fused with the rest of The University of Manchester. The prospectus now shouts *old stone*, a retrograde ivy-league vision that talks of heritage and 1824 instead of the white heat of modern technology. The list of Humanities courses is so very much longer than Engineering and Physical Sciences. And to cap it all, the Jodrell Bank Centre for Astrophysics housed in AMPPS has been threatened by UK science budget cuts. Things have changed, and I don't much like it.

But who am I to complain? Computers were the next big thing when I was starting out. Hence I'm in IT, and it's served me well. I live in a place that people think of as Sandbanks. It's actually called Lilliput, and there really is such a place. It's a comfortable world of beach and pool, fine homes, good schools, ten minutes to work, quality of life. There's time to do the things you're interested in doing, and my interest is physics. It's more than a hobby. It's more of a passion, one that's grown deeper as I've grown older. I'm deeply curious about the world, I want to know how the universe works. Doesn't everybody?

No. *The sound of a stylus slewing across an album.* The wife and I have two teenage children, and when the time arose for choices and courses, I was disappointed. Both the girl then the boy ended up dropping all their science subjects. I asked them about it, talked to them. They told me physics at school was dull. They weren't taking it forward. What can you do? You can lead a horse to water but you can't make it drink. They're simply not interested in science. It's my fault. I wasn't paying enough attention, either to them or their schooling. I didn't notice that the curriculum had changed. I didn't notice how the dead hand of Health & Safety has turned it from Van Der Graaf generators into bookwork, where the most exciting experiment involved timing a pendulum. I'm not sure of the underlying reasons, but whatever they are, it's real, it's happening, and I didn't realise until it was too late. Suddenly I woke up. I learned of physics departments closing down, and other science departments too, and it set my alarm bells ringing. I read that the number of A-level students taking physics had fallen 57% in 20 years, and I felt a shiver. This is serious. *What are you going to do about it?*

<table>
<tr><td colspan="9">The table shows the number and proportion of students taking A-level science subjects, 1984 – 2006</td></tr>
<tr><td></td><td colspan="2">1984</td><td colspan="2">1994</td><td colspan="2">2004</td><td colspan="2">2006</td></tr>
<tr><td></td><td>Number</td><td>%</td><td>Number</td><td>%</td><td>Number</td><td>%</td><td>Number</td><td>%</td></tr>
<tr><td>Biological Sciences</td><td>45,171</td><td>7</td><td>41,489</td><td>7</td><td>44,290</td><td>7</td><td>46,624</td><td>7</td></tr>
<tr><td>Chemistry</td><td>48,068</td><td>8</td><td>34,722</td><td>6</td><td>32,151</td><td>5</td><td>34,534</td><td>5</td></tr>
<tr><td>Physics</td><td>54,722</td><td>9</td><td>30,810</td><td>5</td><td>24,645</td><td>4</td><td>23,657</td><td>3</td></tr>
<tr><td>Other Science</td><td></td><td></td><td>4,415</td><td>1</td><td>3,775</td><td>1</td><td>3,559</td><td>1</td></tr>
</table>

Source: DfES Statistical First Releases.

The number studying physics, once the most popular science A-Level, has fallen by 57% in just over 20 years. Over the same period, take-up of chemistry has dropped 28%.

Figure 3 – Physics A-levels

All this was before the new baby, but that's another story. At the time I was left saying I'd left it too late for my own children, but perhaps I could do my bit to enthuse other people's children. Perhaps I could use my skills to make physics a little more interesting, a little more accessible, and a little more fun. Maybe I could also appeal to the man in the street, and even the sort of person who calls in an electrician to replace the halogen spotlights. I say that because there's a standing joke in our house, that *I'm the only one who can change a light bulb.* But somehow it isn't funny. If you selected a hundred people at random and tested their technical and scientific knowledge, I think the average score would be lower than that of a comparable group from fifty years ago. Yes, we're more specialist these days, and some things are more difficult to understand. But it seems there's more people around who just don't understand the basics, who have only the vaguest concept of how things work. They wouldn't know where to start if their car broke down. It's like there's a low-rise, low-brow tide that doesn't feel healthy, that slowly, insidiously, is getting worse. *Something must be done,* I said to myself. And if you want something doing, you've got to do it yourself.

So there I was, determined to make a difference. I do have a skill. It's subtle, understated, you don't always see it. Like I was saying, I've spent my years in IT, where we work hard to keep things simple. It's all to do with language, the right language. A good system is clear and understandable, organised in a top-down structured fashion so you can look at a picture and get a feel for it quickly.

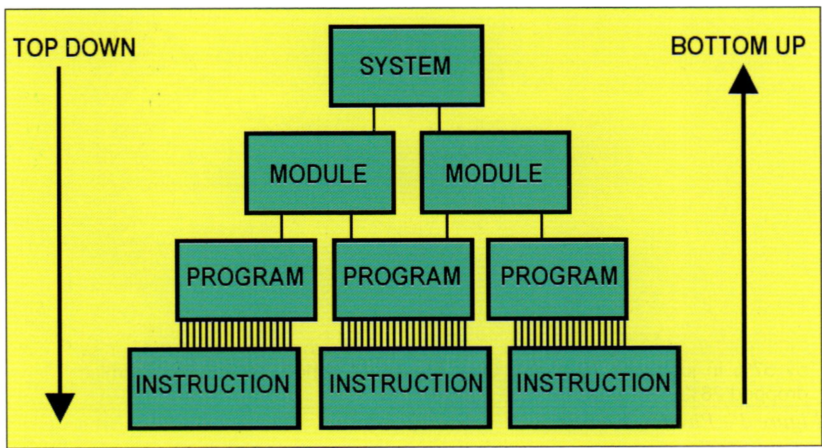

Figure 4 – Top down and bottom up

Then you can dig down a level and get a handle on a few more things, and keep on digging until you get to the bottom of it. Every step of the way you're seeing order, discipline, clean functionality, understanding. It's all so easy when you make it look easy. That's what Systems Analysis is all about. You call a spade a spade and make sure it does what it says on the can. You learn to think clearly and logically, and when you do it right, nobody knows you did it right. But *you* know you did it right, because you've seen what happens when people do it wrong. Because when it's *not* done right, nobody can understand it. That's the trick of it. Nothing under this Sun is beyond the wit of human understanding. When it looks that way, it's because somebody messed up. And when you finally clear it up you wonder how people could get it so wrong.

Where would I start? At the beginning of course, with the basic concepts. I'd explain them simply and logically in a friendly fashion that made physics fun. After all, as Einstein said: *If you can't explain it to your grandmother, you don't understand it yourself.* No problem. I mean, we're talking popular science here, not rocket science.

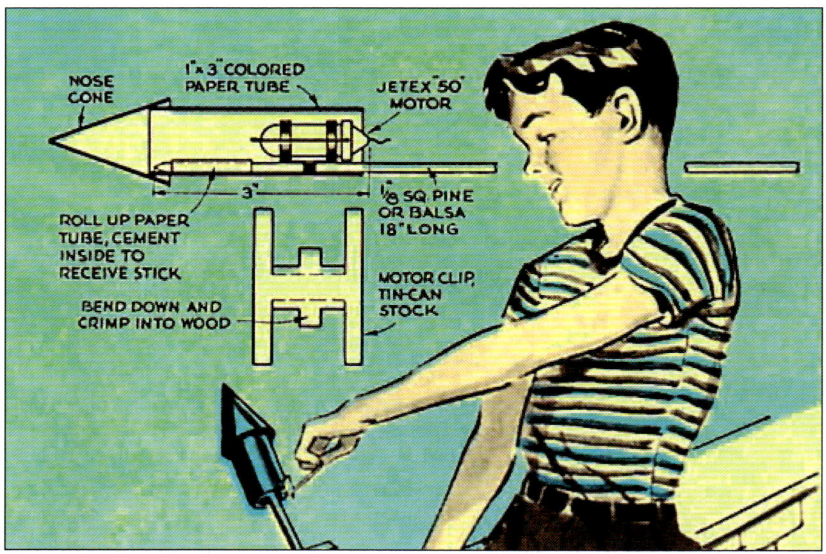

Figure 5 – Popular Science rocket science

No problem? *Big* problem. The problem was that when I tried to explain the basics, I found I couldn't. *I couldn't explain it to my grandmother.* I consulted my books and searched the internet, I delved deep, I read papers, but I just couldn't find the answers. Then I looked at the mathematics, and thought about it. I applied my Systems Analysis experience, I broke down the problem and shuffled the pieces and checked the functionality of every component. I worked it out, I found out what my problem was: mathematics is a vital tool for physics, we can't do physics without it. But the basic concepts I was trying to explain is where the mathematics starts. And whilst I take a top-down approach, the mathematics is *bottom up*. Those basic concepts relate to base mathematical terms, like the E and the m in $E = mc^2$. Mathematics doesn't explain them, because these are the things it uses within its constructions. I was trying to dig under the mathematics, I was analysing the axioms, the self-evident truths, the postulates, the things we take as read, and I realised that when it comes to basic concepts, mathematics doesn't have a handle. It doesn't offer any *grasp*. Don't get me wrong. With mathematics we calculate and predict, and then conduct experiments to confirm these predictions. It really is vital. But mathematics shouldn't be the only tool in the box. It can't do it all, it can't explain the basic concepts. That's like trying to teach English in a foreign language, or trying to understand justice by learning legal shorthand. *You just can't do it.*

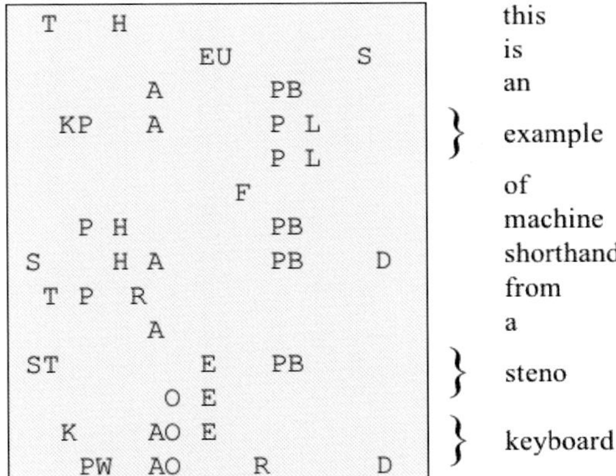

Figure 6 – Stenography

I understood then why relativity and quantum mechanics[3] have not been reconciled. Relativity is macroscopic and top down. Quantum mechanics is microscopic and bottom up. It isn't just a gap between them, they're headed in different directions, and they've missed each other. And to get to the bottom of why, I had to get to the bottom of those basic concepts. That's when the fun began. That's how I got where I am.

I'd like you to come with me on a journey, and enjoy *the pleasure of finding things out*[4]. It's all about basic concepts. Concepts so basic that you've perhaps never really thought about them. But when you do, you realise how irrational some of our ideas are, and how we cling to convictions that have absolutely no supporting evidence whatsoever. It's all about energy, mass, charge, time, gravity, and space, examining these mysteries with slow measured logic in simple conceptual terms. It builds into a mental model that fits with everything you know. It's Einstein's dream, a world of pure marble geometry. This is what he was working on, and never finished. Gödel handed him the key when they were together at the Institute for Advanced Study in Princeton. It's both disturbing and it's wonderful. So elegant and so simple, and we were so close.

What a shame that Richard Feynman, the "Great Explainer", turned down a special teaching and study position at Princeton after the war. He went to Caltech instead[5]. Or alternatively what a shame that Robert "oil drop" Millikan[6] didn't get Einstein to take a permanent position at Caltech instead of going for Princeton in 1933. If Einstein and Feynman had been together, maybe, just maybe, Feynman could have become Einstein's successor, carrying the torch.

Figure 7 – Quantum electrodynamics

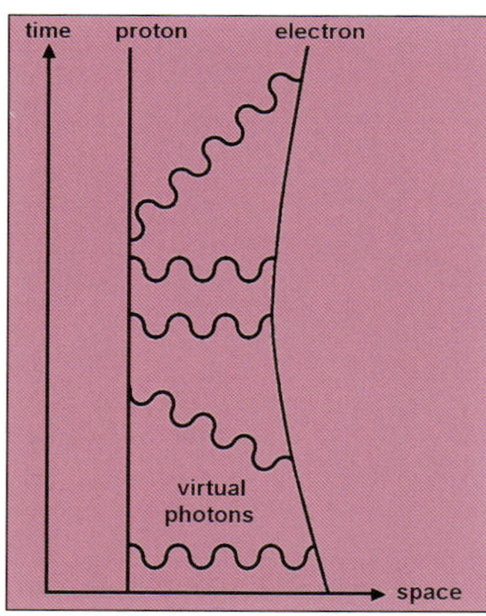

Figure 8 – Feynman diagram

Feynman won joint Nobel Prize for physics in 1965 along with with Sin-Itiro Tomonaga and Julian Schwinger *for their fundamental work in quantum electrodynamics, with deep-ploughing consequences for the physics of elementary particles.* Like Einstein he's my hero, and what makes it so poignant is that the gap between quantum electrodynamics and Einstein's twilight work is so easily bridged.

If only Einstein had somehow passed on what he knew to Feynman. Maybe we would have had a theory of everything by now, something that takes the best of quantum field theory and blends it into a new improved relativity. A relativity that actually explains what gravity is, that tells us how to master it. But it didn't happen. So our rocket science remains complicated, and difficult. Too difficult:

Figure 9 – Challenger disaster

After the Challenger disaster it was Feynman who told NASA they'd been fooling themselves about safety, and about O-rings[7]. All it took was a touch of frost to cripple the space programme. It's incredible how they could get it so wrong.

I think that if Einstein had gone to Caltech or Feynman had gone to Princeton, Feynman would have realised that this applied to other things too. And then things would have turned out different. So different that by now, NASA wouldn't be flying rockets with O-rings. They'd be flying something faster, further, higher. They wouldn't be reaching for Mars, they'd be reaching for the stars. So come with me on that journey. But hold on to your hat, and hang on tight.

Because it's quite a ride.

A BRIEF HISTORY OF RELATIVITY

If you ask for the potted version, you'll be told that Einstein's theory of relativity was developed in two stages. Special relativity came first, giving us the famous $E=mc^2$. It's based on the idea that the laws of physics are the same for all observers regardless of their motion, and that the speed of light is constant. The result is time dilation and length contraction, and the fusion of space and time into spacetime. As you live your life you trace out a world line in this spacetime, always staying within your light cone, and we use Lorentz transformations to work out how space and time rotate into one another for different velocities. General relativity came later, as a development of special relativity rather than a separate theory, telling us that gravity and acceleration are equivalent. It leads on to gravitational time dilation and curved spacetime described by Einstein field equations and metric tensors.

It all sounds perfectly reasonable, but the history of relativity holds some surprises for the general reader. In 1905 Einstein built on the work of Voigt, FitzGerald, Lorentz, and Poincaré to publish a paper called *On the Electro-Dynamics of Moving Bodies*[8]. But he didn't acknowledge his predecessors at all. It's rather controversial, see for example *The Mystery of the Einstein–Poincaré Connection*[9] by Olivier Darrigol. Some even talk of plagiarism, and I wonder if there's some truth to it. Whatever the truth, many agree that Einstein ended up with more credit than he deserved, and others less.

Figure 10 – Woldemar Voigt, George FitzGerald, Hendrik Lorentz, Henri Poincaré

Wherever it came from, Einstein told us there was no absolute time, using a postulate that says the speed of light is always measured to be the same. But note that he didn't say the speed of light is always the same. He said it was always *measured* to be the same. The difference is crucial, and what's also crucial is *why* the speed of light is always measured to be the same. This seemed like an obvious next step for Einstein, particularly with the advent of *Presentism*[10] by John McTaggart in 1908. That was the year relativity became relativity. Until then Einstein had been calling it the theory of invariance, which has an opposite meaning. It was Max Planck, the founder of quantum theory, who called it relativity. Planck spotted it, promoted it, and championed it. If it hadn't been for him we probably wouldn't have heard of it. And 1908 was also the year when Hermann Minkowski invented the concept of spacetime with *Space and Time*[11], so cementing time as the fourth dimension. That was thirteen years after the concept appeared in *The Time Machine* by H G Wells.

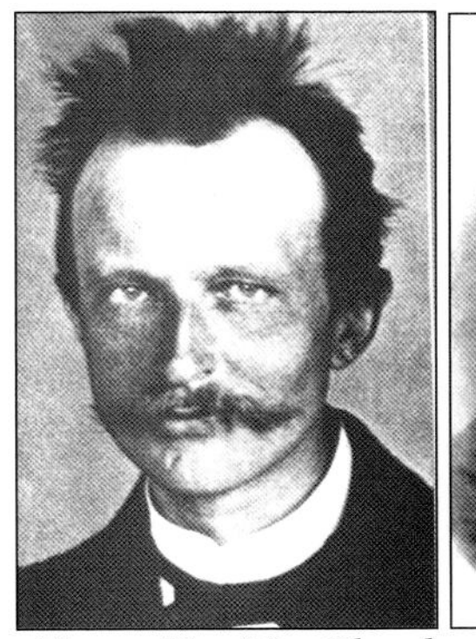

Figure 11 – Max Planck and Hermann Minkowski

Minkowski had been Einstein's maths teacher at college. Einstein used to skip Minkowski's lessons, and Minkowski called Einstein a lazy dog. So it isn't surprising that Einstein was rather scathing. He ridiculed Minkowski by talking of "superfluous erudition", and famously said:

"Since the mathematicians pounced on the relativity theory I no longer understand it myself".[12]

Einstein later acknowledged Minkowski's efforts along with other mathematicians such as Gauss, Riemann, Christoffel, Ricci, Levi-Civita, and Grossmann when *The Foundation of The General Theory of Relativity*[13] was published in 1916. But there was a race or even a priority dispute with David Hilbert, who had been close to Minkowski[14]. And the original translation doesn't employ the phrase "curved spacetime". Nor do you see mention of "world lines". And you have to actually read it to see that it's simply not in accord with the world according to Minkowski. What you read is this:

"We shall soon see that the general theory of relativity cannot adhere to this simple physical interpretation of space and time... The method hitherto employed for laying co-ordinates into the space-time continuum in a definite manner thus breaks down... Let p and ρ be two scalars, the former of which we call the 'pressure', the latter the 'density' of a fluid..."

It's nothing like what you thought it was like. Einstein's talking about pressure and density. There was jockeying going on, rivalry, competition. And when it comes to Minkowski, Einstein seems to have been toeing a line, pretending one thing and saying another. Perhaps it was because Minkowski died in 1909 at the age of only 44. If he'd still been around it might have been easier to criticise his concept of spacetime in those early years. Who knows, but in 1916 Einstein wrote it up in a popular science book entitled *Relativity: The Special and General Theory*[15] and there in the appendix he says:

"We can regard Minkowski's 'world' in a formal manner as a four-dimensional Euclidian space (with imaginary time co-ordinate); the Lorentz transformation corresponds to a 'rotation' of the co-ordinate system of the four dimensional 'world'..."

He really used those quotes around 'world' and 'rotation'. He's telling you the time co-ordinates are imaginary, and spacetime is an abstraction. He's definitely not in line with Minkowski. What you thought he thought, is not what he thought. And he thought some things that really don't square with the way relativity is taught today. Like this from chapter 22 of the same book:

"In the second place our result shows that, according to the general theory of relativity, the law of the constancy of the velocity of light in vacuo, which constitutes one of the two fundamental assumptions in the special theory of relativity and to which we have already frequently referred, cannot claim any unlimited validity. A curvature of rays of light can only take place when the velocity of propagation of light varies with position".

It's crystal clear from the context that when Einstein used the German word geschwindigkeit, which means both velocity and speed, *the constancy of the velocity of light* means he was talking about speed. He was talking about speed because even something as simple as a mirror changes the velocity of a beam of light. When he approved the translation he was thinking of the common usage, as in "high velocity bullet". And yet people cling to a vector-quantity velocity to suit their curved spacetime preconceptions, wherein the change of direction is the change in velocity. This reduces Einstein's words to a ridiculous tautology. He would have been saying *light curves because it changes direction*. His real meaning is obvious, it's clear as a bell even if it's not common knowledge. He's talking about a variable speed of light. There it is in black and white from the man himself: *die Ausbreitungsgeschwindigkeit des Lichtes mit dem Orte variiert*. The speed of light isn't constant, and Einstein said it.

There's more of this sort of thing. Things that absolutely do not square with what people think relativity is. For example in 1920 Einstein gave his Leyden address[16] which included this statement:

"According to the general theory of relativity, space is endowed with physical qualities; in this sense, therefore, there exists an ether. According to the general theory of relativity space without ether is unthinkable; for in such space there not only would be no propagation of light, but also no possibility of existence for standards of space and time".

Einstein was talking about aether, which special relativity had supposedly dispelled. But very few people are aware of this. And very few people are aware that Einstein began to be sidelined from 1922 when he started work on a unified field theory, then split from the mainstream in 1927 over quantum mechanics. And not many people know about Einstein's Princeton work with Gödel in the late 1940s.

Figure 12 – Albert Einstein and Kurt Gödel

Kurt Gödel was the logician who upset mathematical formalism with his Incompleteness Theorem[17] in 1931. Contrary to what people say, Gödel didn't "find a way to time travel" with his rotating universe in 1949. He merely used this conjecture to demonstrate that "time could not have passed if you could visit the past". Here's a interesting quote from a book called *A World Without Time: the forgotten legacy of Gödel and Einstein* by philosopher Palle Yourgrau:

"They walked home together from Princeton's Institute for Advanced Study every day; they shared ideas about physics, philosophy, politics, and the lost world of German-Austrian science in which they had grown up. What is not widely known is that in 1949 Gödel made a remarkable discovery: there exist possible worlds described by the theory of relativity in which time, as we ordinarily understand it, does not exist".[18]

The history tells us that Einstein's ideas weren't quite what we thought, and that they evolved, they changed, they grew. It wasn't as simple as people say. He didn't come up with a theory of relativity in 1905 that matured in 1916 and remained pristine until he died in 1955. And nor has it remained pristine since. Relativity didn't begin to enjoy its "golden age" until 1960, and was then altered in ways that Einstein simply wouldn't agree with. But people appeal to his authority to persuade you to believe what they say, even though they contradict what he said. It's a horrible history. Einstein comes out of it as a tragic hero who won the hearts and minds of the media and public whilst earning stony hearts from his contemporaries. A hero who ended up out of the mainstream, scribbling away on his own trying to finish an important theory, while some thought he was a crazy old fool and others corrupted his memory.

Where was he going in his dying days at Princeton? He was working on a classical unified field theory, but *in his last paper on the subject, admitted that perhaps the concept of field was inadequate*[19]. It was a unified field theory that talked about a displacement field, but it was going to dispense with fields. It sounds bizarre, but it isn't. Not when you understand what an electric field really is, and what gravity really is. I think Einstein worked it out at some point, but couldn't find a way to say it. He was onto the truth, but people couldn't handle the truth. He was conscious of posterity. He didn't want to risk his reputation, he needed rigor. He needed to nail it with mathematics, but he couldn't find a way. And then in 1955 time ran out. Einstein died and nobody picked up the torch. We've been blundering around in the dark ever since, whilst the light of truth has gone begging.

And meanwhile it's taken seventy seven years to solve the mystery of the möbius strip. No wonder Einstein couldn't find a way.

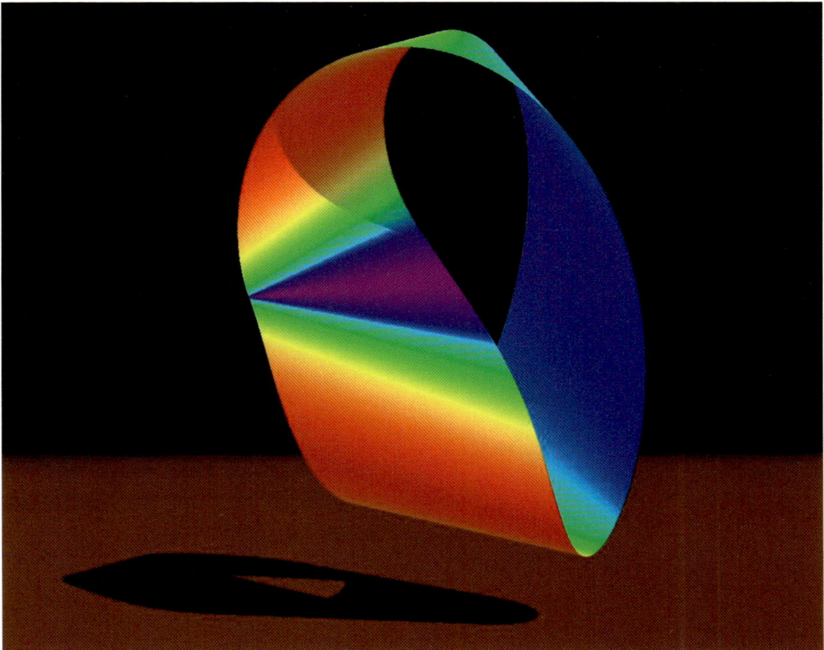

Figure 13 – Möbius strip

You wouldn't believe how simple it all is. It's like riding a bicycle. If you'd never seen a bicycle before, you might think it was impossible. It isn't. It's easy when you know how. It's all to do with $E=mc^2$, all to do with energy and mass, and light and time. And it's all to do with electrodynamics, because that's what Einstein's 1905 paper was all about. Once you under-stand these things, the problems start tumbling like dominoes.

MONEY EXPLAINED

I know it might not sound like physics, but bear with me. I want to give you an example of something that a lot of people take for granted and never really think about. They spend all their lives chasing money, it occupies their thoughts, but they never sit down and think about what it really is.

Show me the money, I say. So you pull out a £10 note. We both know that's money right? Wrong. Check the small print: "I promise to pay the bearer on demand the sum of ten pounds".

Figure 14 – Ten pound note

Your tenner isn't really money. It's what's known in the trade as a promissory note. A mere *promise* to pay money. Basically it's an IOU, from the Bank of England. Of course if you're in the USA or Australia you don't get the small print, but your buck is still a promissory note, a mere IOU. It's just a piece of paper. It's not *really* money.

Figure 15 – Five dollar bill

OK you say. *How about this here penny?* You hand it to me. I turn it over in my hand. It's coppery and shiny. New. Freshly minted. But what is it? It's a piece of stamped metal. Get busy with a hacksaw and you'll find that nowadays they're copper-plated steel, but there's been all sort of variations involving copper and tin and zinc, usually alloyed as a bronze.

Figure 16 – Pennies

I could make them in my garage. But it isn't worth it, especially *"Since May of 2006, all circulation Canadian pennies from 1942 to 1996 have an intrinsic value of over $0.02 USD based on the increasing spot price of copper in the commodity markets"*. A penny is just a disc of shiny metal. That's what it really is. It's similar to the milk tokens I remember from when I was a kid. And those useless slot-machine tokens I brought home from Blackpool. Your shiny new penny is just a glorified milk token, acceptable to more than just the milkman. It's a money token. Notes and coins are *currency*. What they are is *cash*. They're all just money tokens. They aren't *really* money. We have to forget about cash.

Where do you keep your money? I ask. *In the Bank* you reply. *Where in the Bank?* I say. *In the vault* you say. To which I say: *But that's just cash.* You pause for a moment, then change tack and tell me your money isn't in the vault. It's in your account. You've got your salary going into your bank account every month. Whoa. *Where is this account?* I say. *In the bank* you say. *Where in the bank?* I say. *On the computer*, you say. And now we're getting somewhere. Your money is just a bit of ghostly intangible information on a computer.

Before they had computers, your money was "in" a ledger. It was just writing in a book. Ink on a page. The money itself had as much real existence as a bit and a byte. All the ledger ever was, was a glorified bar tab. That's all your bank account is. And all that happens when your salary goes into your account, is that your employer tells your bank to reduce his bar tab and increase yours.

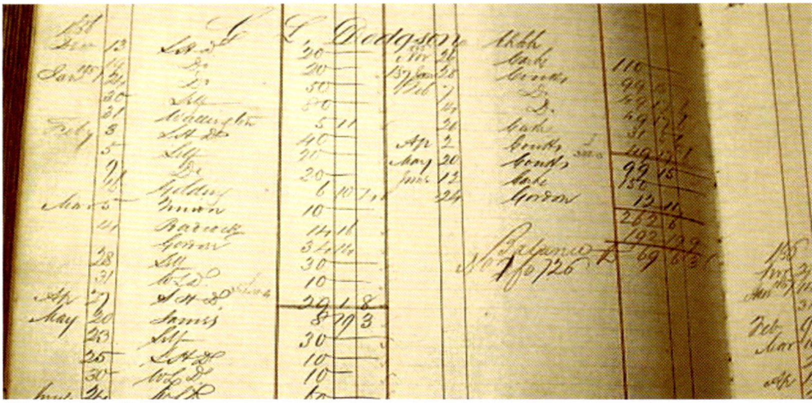

Figure 17 – Ledger

The truth of it is that money *is* just a glorified bar tab. An agreement about IOUs. Nothing is actually moving into anywhere, or out of anywhere else. *Ah*, I can hear you saying, *what about the gold standard?* Shrug. Copper, gold, what's the difference? Gold was only "money" because everybody agreed that this nice shiny metal was desirable. So were pretty little sea shells once upon a time. They were money too. Imagine a pirate landing on a deserted island, the native people wiped out by some pestilence. The pirate kicks amongst the ruins, shells crunching underfoot, hawking and spitting and sick as a parrot at finding no gold.

Figure 18 – Pirate

Because there *isn't* any money if people don't *agree* that it's money. Because money doesn't exist. Not really. That's why when you spend money it doesn't disappear. It isn't destroyed. You've got less of it, and the shop's got more, but nobody's really got more or less of anything.

Do you know how money is created? Banks lend money to people who build houses and cars and flatscreen TVs. Then banks lend more money to people who buy those houses and cars and flatscreen TVs. And then the money that was magicked up on a computer becomes money for real. Because the things exist, and the value exists, and everybody agrees on that value. But the money is only agreement, it's only notionally there because people agree on value. It doesn't really exist, so it never really gets created or destroyed. Yet it gets things done, and it makes the world go round, and everybody wants it. And all it really is, is a glorified bar tab. It's merely a tokenised agreement of value that we label as "a medium of exchange". Then we take it for granted, and think it's the most important thing in the world.

For a lot of people, realising what money really is comes as quite a surprise. It's obvious when you think about it, but people usually don't. There are things in physics like this. People talk about them without thinking about what they really are. But once you do, that's when doors start opening.

THE PSYCHOLOGY OF BELIEF

When I analysed my basic concepts, I found things that weren't real, that don't exist, that we never actually see. But we assume they're real, we take them for granted, and we believe in them. Because we have holes in our understanding, holes that we've all grown up with. We've lived with them for so long that we don't know they're there. We cover them up with ignorance of our ignorance, with blindness of our blind spot, and we shield ourselves with a peer pressure that persuades us there are no alternatives to consider. We do it because we are social animals, we follow the herd, we're prey to groupthink. That's the way we are. So much so, that we even place our faith in negative carpets.

What's a negative carpet? Well, let's say that the wife is so impressed with the new lounge carpet, that she now wants a new carpet for the baby's bedroom. The room is square, and we need sixteen square metres. What's the square root of sixteen? There are two solutions, four and minus four. So wise guy that I am, I opt for the latter solution, and get down on my hands and knees to cut a big fat square out of our brand new living room carpet. I roll it up, put it over my shoulder, and take it to the carpet shop, walking backwards for dramatic effect. I hand it over to the proprietor and pay him a minus ten pound note, which I stick in my pocket, then go back home to crack open a bottle of wine and greet my guests. We're standing in the living room talking about my negative carpet and discussing its negative mass when the wife walks in. She stands there open-mouthed for a heartbeat or two as I begin to explain the merely technical details of relocation to the baby's bedroom. Then all hell breaks loose.

Figure 19 – Hell

The thing about all this, is that a solution is sometimes crazy, but it's not always plain. People just don't spot it. So we talk about it quite seriously without examining whether it's a real solution. We end up taking it for granted and using it to search for further solutions. Then when we struggle, we forget to track back to the beginning and look at the things we took for granted. We don't realise we're riding a negative carpet to never-never land, and that's why we're getting nowhere. What it all boils down to, is that a negative carpet doesn't exist. It isn't real. It's just a figment of our imagination, an abstraction, a *belief.* And beliefs can cause all sorts of problems. Some people believe in Santa Claus, and some people believe in fairies, despite that fact that there is absolutely no material evidence to support the existence of these things. We smile at the gullibility that foolish people show, but we forget that we too believe in things for which there is no material evidence. Things like time travel, unseen dimensions, and parallel worlds:

Figure 20 – Parallel worlds

We've all got our beliefs. That's the way we are. I've got them, and so do you. It was Feynman who said *"The first principle is that you must not fool yourself, and you are the easiest person to fool"*[20]. This is more true than you realise. It's true because when you've fooled yourself, *you don't know it*. You convince yourself that you haven't fooled yourself, and you develop a conviction, a faith, a *belief* about it. You'll be quite irrational in defence of this belief. You won't test your belief in an empirical scientific fashion. Instead, when challenged, you'll become defensive or incredulous. If you don't behave this way, that's fine, you're not a believer. You merely have an opinion, and an open mind. But let me demonstrate something: *You don't have an open mind. You're fooling yourself.* At which point I imagine you're bristling already. See how it works? If you really believe something and I challenge it, it's all too easy to construe the challenge as an insult, and then become hostile and unreasonable. That's human nature. *Everybody* likes to think they have an open mind, and very few understand that about some things at least, they don't. The truth is this: you're not quite as open minded or as rational as you think. This is hard to accept, but that's the way it is. It's like that because if you *believe* something, you don't *need* to think about it. Because you already *know* the answer. Hence you're less receptive than you should be. And so you don't look at the out-of-the-box solutions that solve the problems that have troubled you all your life.

Stop a minute and think about it. Why do you think we have suicide bombers? What on Earth possesses them to think that there's seventy two virgins waiting for them in paradise? What possesses them is something called *The Psychology of Belief* [21]. And they *don't think*, that's just it. This thing is far more powerful and far more prevalent than you know. There's a whole spectrum of belief out there. Think about Young Earth Creationists and their Intelligent Design friends. You can talk to these people until you're blue in the face, but they're totally immune to logic because they *believe* that they're right. You can say anything and everything, but they duck and dive and dismiss every last scrap of evidence you throw at them. Everything you say goes *whoosh*, in one ear and out the other. They just aren't listening. They just aren't thinking. The weird thing is that they don't know they're immune to logic. These guys aren't lying to you. They don't have a rational open mind, but *they don't know it*. They think they're being perfectly rational, and you're just some crazy fool who just doesn't know.

Figure 21 – Intelligent Design

It doesn't stop at religion. There's ideology, Kafkaesque bureaucracy, and dynastic communism, all the sorts of things that can end up with starvation, murder, and Nazi death camps. There's racism, tribalism, and insane conspiracy theories, all leading to enmity and hate and violence. There's heroin, crack, and alcohol addiction where people die before their time. Moving down the scale there's anorexia and obesity, and the dieting that makes you fat as your body sets store for a rainy day. Then there's gentler symptoms like fashion, where folk let themselves be brainwashed into thinking purple is the new black. Or swaggering around with some eco cotton bag containing the keys for the 4x4 and the plane tickets. It affects everybody to some degree, even people who consider themselves to be utterly rational and totally open minded. Everybody's got some kind of belief about something. When you find it and hit it, *whoosh*, everything you say goes in one ear and out the other. They just don't listen. They just don't think. It's like the shutters are down and there's nobody home.

Would you like to put yourself to the test? This will show you what I mean. This will demonstrate that you're not immune to *The Psychology of Belief*. Nobody is, not even me. Take a look at the picture below:

Figure 22 – Checker-shadow illusion by Edward Adelson

OK, here's the deal: *squares A and B are the same colour*. They're the same shade of grey. *Oh no they're not*, I hear you say. *Oh yes they are* I insist. *Oh no they're not* you answer back. We could do this all day, but I'm afraid I'm right and you're wrong. They really *are* the same colour. Squares A and B are the same shade of grey. The apparent *difference* in colour is the illusion. Let me prove it. It's very simple. Just tilt the page so you're looking at it from a narrow angle. Alternatively fold the paper to get the two squares next to one another. Another method is look through a small hole to remove the context that fooled you into fooling yourself. You can look at it online if you wish. Google on "checkershadow illusion" or go directly to <u>http://web. mit.edu/persci/people/adelson/checkershadow_illusion.html</u>, then you can see for yourself. You can even download an image and check it out with photoshop. Satisfy yourself. Be empirical, test yourself, find a way to stop fooling yourself. Then you realise that A and B really *are* the same colour.

Figure 23 – Checker-shadow illusion revealed

Don't be surprised. I told you *The Psychology of Belief* is powerful. More powerful than you ever dreamed. What's surprising is just how common it is, even amongst scientists. If you don't believe me, you should look up *paradigm* on Wikipedia, and read *The Structure of Scientific Revolutions* by

Thomas Kuhn[22]. Conviction is a hard nut to crack, and it applies to scientists too. It's the way we are, the way we think. Why do you think Bruno got burned at the stake? Why do you think it took Einstein seventeen years to get a Nobel Prize for the wrong thing? And why do you think there's that saying: catch 'em young? It's because there are people out there who are quite fully aware that if you instil children with a belief they'll carry on believing it come heaven or high water. These children remain so utterly convinced, that they grow up to become adults who will fight and *die* for it. But we're not going to fight and die for something like *The Capacity To Do Work* are we? Because we are rational, we have an open mind, and we listen and we think.

Yes, *The Capacity To Do Work*. Einstein said you don't really understand something unless you can explain it to your grandmother. Can you explain energy to your grandmother? You might believe you can, but the chances are you're fooling yourself, and your explanation is no explanation at all. Your grandmother will peer at you over her bifocals, suck on her false teeth, say *Thank you Dear*, and then she'll carry on with her knitting. She's too polite to say it, because butter wouldn't melt in her mouth. But what she really meant is: *Capacity To Do Work my arse*.

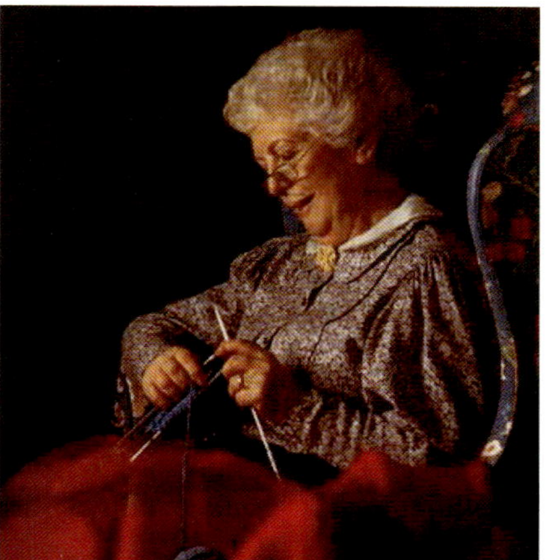

Figure 24 – Explain it to your grandmother

Come on now, *The Capacity To Do Work* is no explanation at all. You swallowed that when you were young and gullible, and you haven't looked at it since. Energy is a simple basic concept that you really ought to understand, but you don't. And you don't know that you don't. Because

Donald Rumsfeld was right[23]. And what you also don't know, is that *The Capacity To Do Work* is merely a label that covers up a hole in your understanding. A hole that you've grown up with, that's been there so long it's like a blind spot, all grown over with such thick skin that you don't even know it's there any more.

I'll show you the holes in your understanding. I'll peel back those labels and fill the holes with concepts that are crystal clear. Then you can stop fooling yourself. But remember this, it's important: *the basic concepts I will give you are better than the concepts you hold now. But don't ever think they're perfect.* Don't fool yourself that you've stopped fooling yourself. Keep that open mind open.

Buckle up. Here we go.

ENERGY EXPLAINED

The schoolroom textbooks told you that energy is *The Capacity to do Work*, and work is the transfer of energy. The words go round in circles without getting to the heart of it, and children grow into adults with no real concept of what energy is. So what is it?

We usually think of energy as being the property of a thing, something you can take away from a thing. But it isn't like that. To illustrate what I mean, I can talk about a red balloon, a red bus, or a red red ruby. All these things have the property that we call red. A thing can be red, but you can't remove this red and hold it in the palm of your hand. You can remove the paint or the dye, and you can hold that in the palm of your hand, but you're still holding a thing that is red. You can remove a thing from a thing, but removing a property isn't always so easy. You can't remove the red from the dye to hold the red in the palm of your hand. Even when you imagine red, the image in your mind's eye is a thing. You always need a *thing* to *be* red. There *is* no such thing as "raw red".

Figure 25 – Red balloon, red bus, and red red ruby

The interesting thing about colour, is that buses aren't really red. The photons of light that reflect off the paint make them look red. Only those photons aren't really red either. They only *look* red, because that's the way you see them. They don't have a colour, they have a frequency. They have a higher frequency when they have more energy, and they have more energy because something gave them that energy.

How do you give something some energy? Think about an old house, nestled in the countryside. It's picturesque, worth a lot of money, and it's built out of cob. Way back when, some medieval construction crew put some energy into shifting earth and straw to make the walls of this house. They did the same with the wood, which grew out of the earth because the trees put energy into shifting water and CO_2. The guys moved stuff, because energy is all about *motion*. But the stuff they moved was heavy. It has mass. So we need to talk about mass and motion to talk about energy.

Let's have a little *gedankenexperiment,* a thought experiment. Consider a 10 kilogram cannonball, in space, travelling at 1000 metres per second. We talk about how much kinetic energy this cannonball has. We say $KE=\frac{1}{2}mv^2$ and we do the maths and get five million Joules. But what has the cannonball really got? Its mass seems real enough, I hefted it into my spaceship this morning before I took off. And its motion seems real enough too, because one false move and it'll be smashing through my viewscreen taking my head off.

Figure 26 – Cannonball in space

To find out more, I take a spacewalk to place a thousand sheets of cardboard in the path of my cannonball. Each sheet of cardboard exerts a small braking force, slowing the cannonball to a halt. This takes two seconds. We know that the cannonball will punch through more cardboard in the first second than in the second second, because it's slowing down. So we reason that a cannonball travelling at 1000m/s has more than twice the kinetic energy of one travelling at 500m/s. We can do the arithmetic for each second, then slice the seconds up finer and finer, and we end up realising that the $\frac{1}{2}v^2$ is the integral of all the velocities between v and 0. But what we don't realise, is that kinetic energy is a way of describing the *stopping distance* for a given force applied to a given mass moving at a given velocity. You can flip it around to think about force times distance to get something moving. Or you can think in terms of damage or work. But basically that cannonball has "got" kinetic energy like it's "got" stopping distance.

It's similar with momentum. That's a different way of looking at mass in motion, based on force and time instead of force and distance. We look back to our cannonball and cardboard, and we know by definition that the same amount of time elapsed in the first second as in the second second. So we realise that a cannonball travelling at 1000m/s has twice the momentum of one travelling at 500m/s. But what we don't realise, is that momentum is a way of describing the *stopping time* for a given force applied to a given mass moving at a given velocity. A cannonball has "got" momentum like it's "got" stopping time.

Now we can apply a little relativity: motion is relative. And I made a deliberate mistake when I told you about that cannonball. Because I didn't fire the cannonball at 1000 metres a second. I dropped it off at a handy spot out near a GPS satellite, then zipped off in my spaceship in a big fat loop.

Figure 27 – Gravity spaceship

It's *me* doing 1000m/s, not the cannonball. The cannonball is just sitting there in space. *It* hasn't got any kinetic energy at all. *I've* got it. But I don't feel supercharged with five million Joules of energy coursing though my veins. So where is it? Where's the kinetic energy gone? It hasn't gone *anywhere* really, because all that cannonball *has* got, is its mass, and its motion. And that motion is *relative* to me. Kinetic energy and momentum are not really properties that a material object has "got". They're just distance-based and time-based measures associated with motion. The motion is relative, and in a way the *motion* has "got" the kinetic energy and the momentum.

So let's look a little closer at motion. How do you make something move? Easy. Hit it with something else that moves. And how did you make that something else move? Where did it all start? I pitch you a baseball, you whack it with a bat, and it flies away at twenty metres per second. You made that baseball move. Now, where did the energy come from to make it move? From your muscles:

"ATP binds myosin, allowing it to release actin and be in the weak binding state (a lack of ATP makes this step impossible, resulting in the rigor state characteristic of rigor mortis). The myosin then hydrolyzes the ATP and uses the energy to move into the "cocked back" conformation. In general, evidence (predicted and in vivo) indicates that each skeletal muscle myosin head moves 10-12 nm each power stroke..." [24]

There's more to it than that, but basically it's all to do with binding, and bond angles. Bonds within molecules change, and the change releases energy. Sometimes it's a simple change of bond angle, something like a leaf spring letting go. Sometimes there's more than one bond angle change, in a molecule that resembles an elasticated deckchair surging from one configuration to another. And sometimes there's a whole host of bonds in play, taking a rather different shape. A familiar shape:

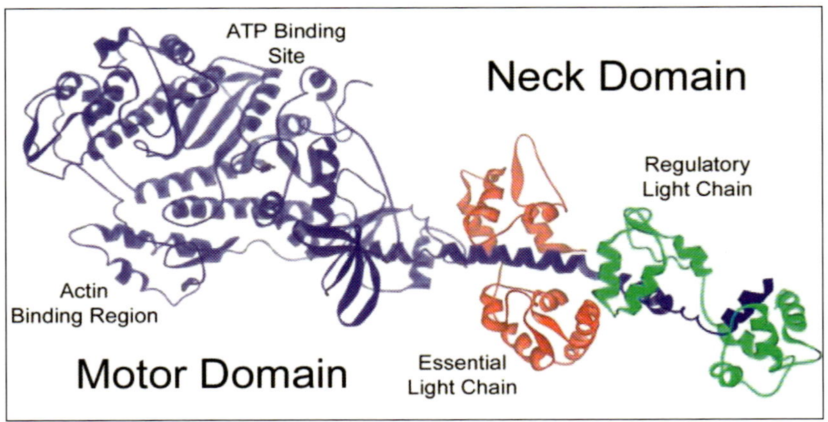

Figure 28 – Myosin structure

The myosin in your muscles looks like a mess of helical springs because that's essentially what it is. Your muscles contract using tiny clockwork motors being repeatedly wound and released via a process called ATP hydrolysis. The molecules look like springs because they *are* springs. That's the size of it. And these molecular springs aren't just in your muscles, they're in your skin too. Your skin is springy because of the elastin. And what does elastin look like? Like this:

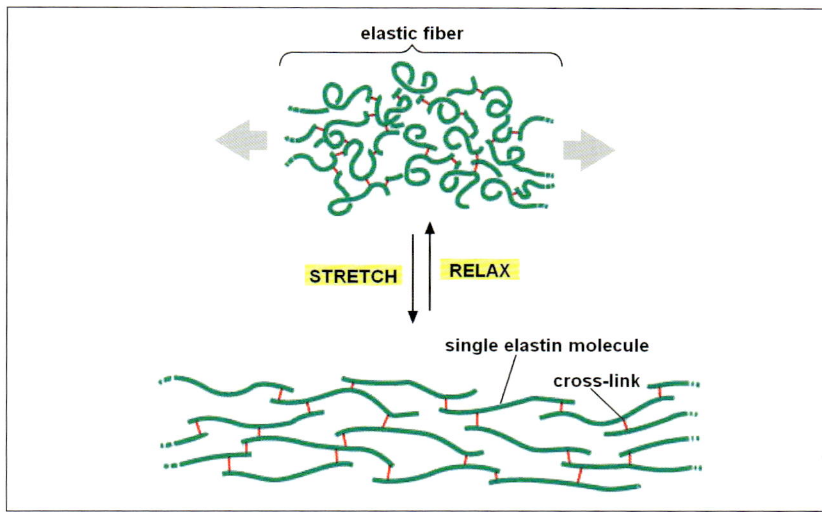

Figure 29 – Elastin

Elastin looks like a heap of interconnected leaf springs because that's exactly what it is. When stretched it's under tension, and wants to shorten. The energy that pulls your elastic skin back in place is stored in springs that are built into the molecules of your tissues. When you get down to the atomic level, all the atoms in those molecules are held together by bonds, and these bonds rely on electrons and the electromagnetic force. They're rather like magnets that want to pull together or push apart, only it's electrostatic attraction and repulsion rather than the magnetic variety. This is how energy is *always* stored in springs, be they molecular or macroscopic, plastic or steel. It's the same for chemical energy. Ever heard of an explosive called cubane?

"Cubane, the hydrocarbon $(CH)_8$, is named appropriately; its skeleton is in the shape of a cube. At each corner of this cube there is a carbon atom (carrying a hydrogen) bound to three identical neighboring carbons… The large bond angle deformations in cubane make it a powerhouse of stored energy. Each strained bond is like the spring in a set mousetrap…" [25]

Did you catch that? It's *strained*. It's under stress. *Each strained bond is like the spring in a set mousetrap.* Explosives are made up of molecular springs, only cubane is stable whilst nitroglycerine has a hairpin trigger.

It's similar with nuclear energy, only now we're talking about the stuff inside the atoms rather than between the atoms, and these springs are stronger. That's why we talk about the strong force rather than the electromagnetic force, but it's the same difference. The Sun gets its energy from nuclear fusion. Squeeze hydrogen atoms together and you make helium. But when you do, *twang*, something lets go, there's a material change, and things spring out between your fingers, things like neutrinos and photons. The recipe goes like this:

$$4 \text{ H} + 2 \text{ e} \rightarrow \text{He} + 2 \text{ neutrinos} + 6 \text{ photons.}$$

As it happens it's a multi-stage process, and we talk about the hydrogen in terms of protons. Two protons fuse together releasing a positron and a neutrino, then a third proton is added releasing a photon, then two trios combine to make a helium atom with two protons left over:

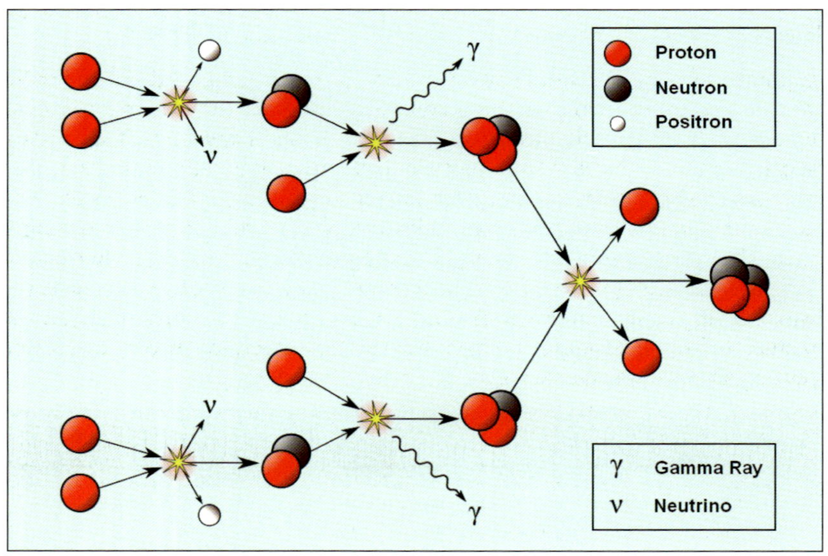

Figure 30 – Hydrogen fusion

The squiggles in the picture above are the photons. Photons are electro-magnetic radiation. The γ means gamma: these are very high frequency *gamma* photons, like X-rays only worse. There's only two shown above, but the two positrons will annihilate with two electrons to make four more photons. The photons might look a little like springs, and you might be thinking *Spring Theory*, but they're not. Because when a positron annihilates with an electron, you're left with no springs at all. Take one positron, add one electron, and *BANG,* matter/antimatter annihilation occurs. The electro-magnetic springs let go so totally, they're just not there any more. It's a total material change, with no material left. The matter is totally converted to energy, and the only product is electromagnetic radiation.

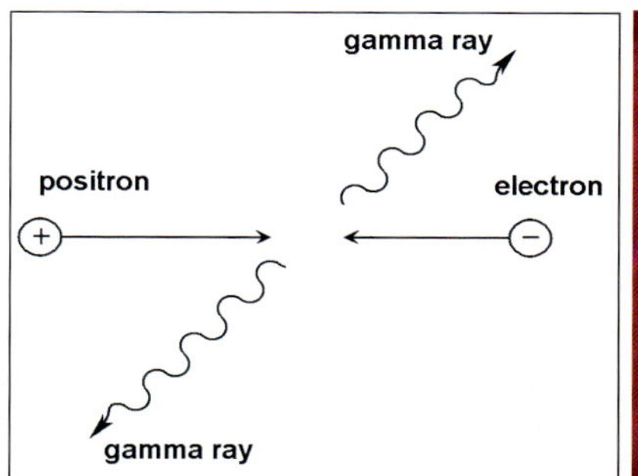

Figure 31 – Annihilation schematic   *Figure 32 – Annihilation animation*

Gamma rays are electromagnetic radiation, so is light, and so are radio waves. Electromagnetic radiation consists of photons, and the photon is of crucial interest. Particle physics comes with mental baggage that says it's a speck, a point, a "billiard ball" particle. But long-wave radio reminds us that photons can be 1500m long, or even longer. A photon isn't some speck. Nor is it some kind of spring. It's more like a ripple in a rubber mat, or the slink in a slinky spring rather than the spring itself. It's a wave, but not a longitudinal compression wave like a sound wave. A photon is a transverse wave. It ripples up and down like this ~ . You can see the difference on the picture below. The upper portion shows a compression wave, the lower portion shows a transverse wave.

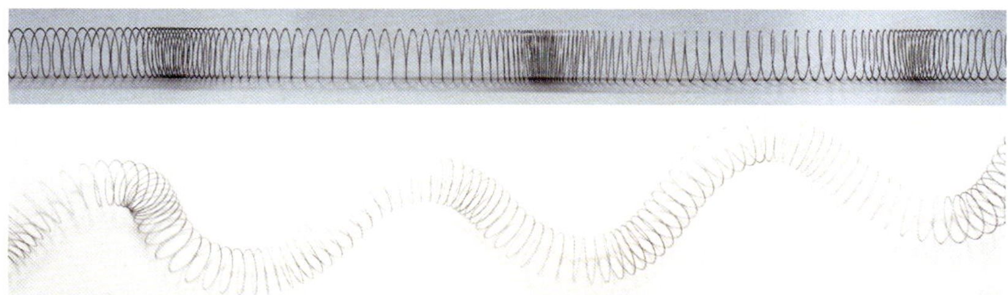

Figure 33 – Slinky

Photons travel through space, that vacuum void with the properties we call electrical permittivity and magnetic permeability. A photon is like a ripple on an electromagnetic ocean between the stars. A boat on an ocean can ride a ripple and the ripple just passes on by. But tie that boat to the sea bed with an elastic rope, and you can capture the energy of the ripple in a bond, just like a plant can capture the energy of sunlight and save it in starch or oil.

Space isn't really an ocean. It doesn't have a surface, it isn't a liquid, and there's no substance to it. But the analogy is fair. A photon is a quantum of "matter/energy stress". It travels in the mysterious fabric of space like a shimmy shooting through a ghostly block of transparent rubber. It's a transverse wave of travelling stress, and this stress is somehow in space itself. It's in the space between the stars, within molecules and atoms, and is everywhere. There is nowhere in this universe where there is no space, and there is nowhere in this universe where there is no energy, so space is always stressed.

Figure 34 – Space

In physics, stress is the opposite of tension, and is the same thing as pressure. To quantify energy and obtain the correct dimensionality, we have to know that stress is force per unit area, and energy is force multiplied by distance. We have to multiply a stress by an area and then by a distance to get energy. So we have to multiply a stress by a volume. We have to multiply the degree of stress by the volume of space that is stressed. Then we can quantify how much energy is there.

There's nothing more fundamental than space. Space isn't made of anything. We can't examine it through a microscope or annihilate it to obtain something else. But we know that space always has a volume. If it didn't, it wouldn't be space. And we know that space is always stressed, because energy pervades the universe. So there's nowhere else to go. It's the end of the line. But it's quite enough for a new definition of what energy really is, and as we shall see, this new definition is utterly profound:

In barest essence energy is a volume of stressed space.

This is why you can't hold energy in the palm of your hand. Because you can't hold a volume of stressed space in the palm of your hand. Just as you can't hold a photon in the palm of your hand, or a sunbeam. Nothing is "pure energy", just as there's no such thing as pure pressure. Because energy is the property of a thing, even when it's the very last property that makes it the thing that it is.

But oddly enough, you *can* hold energy *in* your hand. It's a subtle difference, but it's very simple. Just squeeze a fist. Use your right hand. Squeeze it tight. Now touch your left thumb to your right thumb. Feel that increased blood pressure. Now look at the volume of your fist. Stress is the same thing as pressure, and there's a volume of it in that fist. Your fist now has more energy. And if you swing that fist, it has even more. As to how, that brings us back to mass in motion. It's all to do with pushing little circles into little helixes, like stretching a spring. And to explain that, I'll have to explain *mass*.

Figure 35 – Fist

MASS EXPLAINED

You know that energy is an intangible thing. You can't hold energy in the palm of your hand, because energy is to do with stress, which is the same as pressure, and you need a volume of stress to get the dimensionality right.

You know that mass is a tangible thing. You can hold an object in your hand and feel the mass of it. You even know that $E=mc^2$, and that the intangible thing called energy can be used to make the tangible thing called mass. But you don't know how, because you don't know how simple it really is.

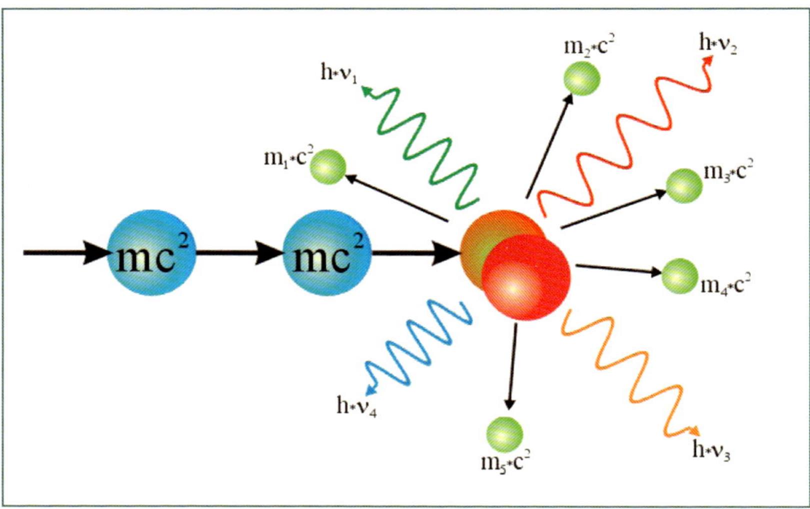

Figure 36 – Energy/mass conversion

The answer is all down to motion. Or the lack of it. You have to get relative, because motion is relative, and you have to think in terms of momentum and inertia. A 10kg cannonball is a tangible thing, you can hold it in the palm of your hand and feel the mass of it. And if it's travelling at 1m/s it's "got" kinetic energy as quantified by $KE=\frac{1}{2}mv^2$. It's also "got" momentum, as quantified by $p=mv$, and it's hard to stop. Brace yourself, then apply some constant braking force by catching it in the midriff. *Ooof*, and you feel the energy/momentum. Kinetic energy is looking at this in terms of stopping distance, whilst momentum is looking at it in terms of stopping time. The measure we call momentum is conserved in the collision because your gut and the cannonball shared a mutual force for the same period of time. The measure we call kinetic energy isn't conserved, because some of the mass in motion is redirected into your deformation and heat and bruises, all of which involve mass in motion, but scattered motion instead of the tidy motion of a mass moving relative to you. Or you moving relative to it, because all the while you were never too sure whether it was you moving or the cannonball.

When we consider that the cannonball is stationary and it's *you* moving at 1m/s, the cannonball hasn't got any kinetic energy or momentum, all it's got is inertia. You're attempting to move it and accelerate it to your velocity of 1m/s, and that makes it harder to start, not harder to stop. In this respect momentum and inertia can be considered as two different aspects of the same thing, and that thing is energy.

You can get a better feel for this with a gyroscope. Waggle it back and forth. See how insubstantial it feels. Now wind the string round the spindle, grasp it tight, and pull. You give the gyroscope some circular motion, you are in effect pulling tension out, and since tension is negative pressure and pressure is stress, you are introducing stress. Your gyroscope is now humming, maybe precessing a little. When you try to waggle it you can feel the angular momentum working against you. It feels more substantial than before. You can feel more inertia, and that's the secret of $E=mc^2$. The gyroscope feels more "massive" because you have added energy.

Figure 37 – Gyroscope

You're beginning to get a feel for mass, but to really understand it you have to stop thinking of momentum as something that a mass has got. A thing that's got energy can have momentum without having any mass. Like a photon. A photon has energy and momentum but it has no mass. And a photon has a size but it has no surface. You might therefore think a photon is something really strange, but it isn't. You've seen something similar down on the beach. You're playing in the surf and along comes a massive wave. You know it's a wave, a travelling stress, and you know it has no mass because the water has the mass. It's a pressure wave, you can see how big it is, and you know it has no surface because the ocean has the surface. But the wave *does* have energy/momentum, enough to knock you and your girlfriend flat on your back, laughing and screaming with salt water up your nose. It's both insubstantial and substantial, both tangible and intangible. Because whilst you can't grab hold of it, *it can grab hold of you*.

Figure 38 – Wave in the surf

When we think about the photon some more, we know that it isn't some miniature cannonball. It's a transverse wave in space, it's got a wavelength and a frequency instead of a mass, so instead of $KE=\frac{1}{2}mv^2$ and $p=mv$ we express energy as hf and momentum as hf/c. The h is Planck's constant of 6.63×10^{-34} Joule-seconds, and is an "action", which is energy multiplied by time, or alternatively momentum multiplied by distance. The f is frequency per second, and the speed of light c is distance divided by time, converting a stopping-distance measure of energy into a stopping-time measure of momentum.

The use of the word "action" is important here. It means what it says, and should doubly remind us that the photon is not some billiard-ball particle. It truly is an action. It's akin to a kick. In itself it's intangible, but like a kick it delivers a quite tangible effect. We can see this in action via Compton scattering, which makes the sky look blue:

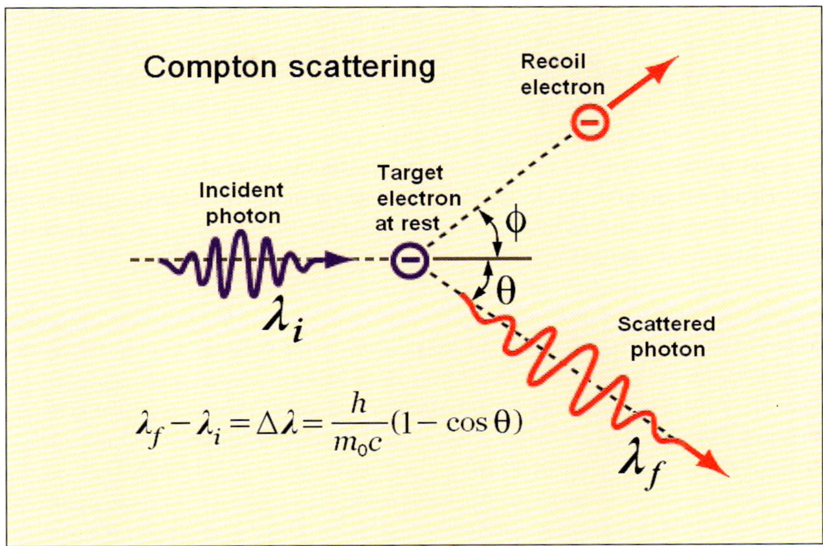

Figure 39 – Compton scattering

When a photon interacts with a free electron, the electron receives a "bump" and is sent recoiling off at an angle, while the photon is deflected and its wavelength is increased. The underlying reason for this is easy to grasp. Look at the central portion of the squiggle that is the incident photon in the picture above, turn it upside down, and imagine it's a U-shaped steel bar. Now grasp it tight and unbend it a little. This increases the distance between the two ends and changes the angle between them. In similar vein the photon wavelength increases and the direction of propagation is altered whilst the energy decreases. Meanwhile the electron gains energy and receives an alteration to its velocity vector, or in other words: it moves.

Now remember your relativity: there is no absolute motion. Look again at the picture. Imagine you're that target electron, but you're not at rest. Imagine it's *you* moving instead of the photon. *Bump*, and you're sent flying off at an angle. It would feel like you hit something solid instead of a photon. It would feel like a bad flight with turbulence and so many air pockets it's like riding over rocks. Instead of delivering a bump, the photon would *be* a bump. It wouldn't feel like something intangible and insubstantial, it would feel like something tangible and substantial. It would feel like the photon had inertia instead of momentum. It would feel like the photon had *mass*.

But a photon doesn't have mass. A photon isn't just sitting in one place, it always travels at c. You can't nail it down like you can nail down a ripple in a rubber mat. So how do you keep a travelling volume of stressed space in the same place? It's easy when you know how. Imagine you've got a couple of "free electron" table tennis bats, and you're good at topspin. If you bat that photon just right, you can change its direction and give it some more energy/momentum. It's called an Inverse Compton, like the picture of Compton scattering but with the arrows going the other way. Then you hit the photon with the other bat to change its direction again. Repeat in rapid succession until you've got a kind of hexagon going, then bat even faster until you've got a miniature electromagnetic swirl that is your photon going round in circles like this ↻. Now keep chopping away, but close your eyes, like you might close your eyes when you're feeling the repulsion between a couple of magnets. You can feel something there between your bats. What you can feel is basically mass. You've "stopped" the photon and converted the momentum into inertia. You've made a "mass".

OK, you haven't really stopped the photon because it's still travelling at c, but it's close enough. *Bah,* I hear you say, *there are no table tennis bats in particle physics. And a photon always travels at 299,792,458 m/s. You can't really "stop" a photon.* Oh yes you can. It's simple. You use something called *pair production*:

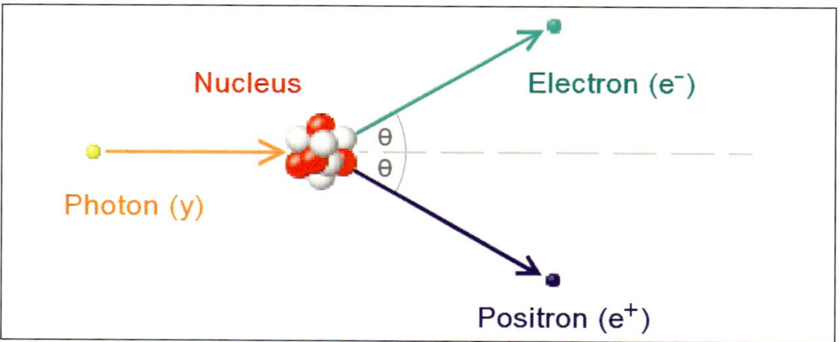

Figure 40 – Pair production

In pair production, a gamma photon of a little over 1,022 thousand electron volts (keV) is effectively broken over a nucleus to create an electron and a positron of 511 keV apiece. Apart from some essential losses associated with the motion and the separation of the particles, the energy/momentum is reconfigured as mass. Pair production converts travelling kinetic energy or relativistic mass into non-travelling energy or rest mass. In effect the momentum is "stopped" and now appears as inertia, and the electron and the positron are like two stable eddies spinning off in opposite directions. Each can be modelled as a self-contained vortex, or vorton[26]. Each is a photon going round and round in a circle, tied into something called a toroidal soliton[27]. It isn't just a circle, it's a circle with a twist, and it needs the twist to stay tied up. This twist makes it a knot, what's called a "trivial knot", which is the simplest knot you can get. An electron twists and turns one way, the positron twists and turns the other, so they're mirror images like this ↺ and this ↻, with different "handedness", also known as chirality. You can think in terms of a *möbius doughnut*, as depicted on the left below. Note the similarity between the dark line as compared to the balloon knot on the right.

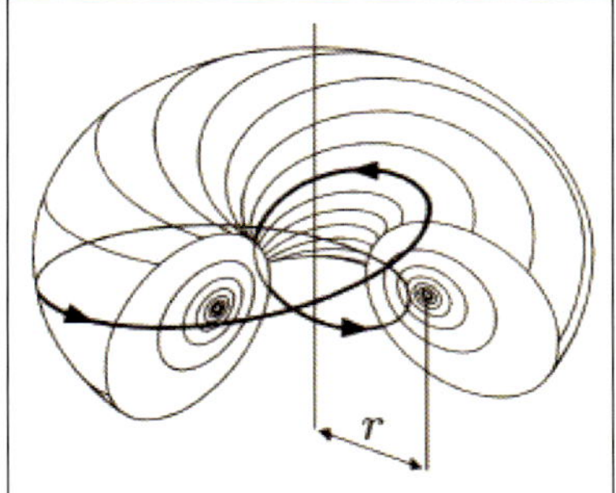

Figure 41 – Toroidal photon electron   Figure 42 – Balloon knot

It's rather like a möbius strip. The electron's component photon is going round twice whilst twisting round once to return to its starting position and orientation. The electron is therefore a spin ½ particle exhibiting a jitter or *zitterbewegung*. It's both a transverse wave and a particle, and hence it's a soliton and hence it exhibits wave/particle duality. It is however somewhat difficult to depict. It isn't a solid object, and whilst we can draw a representation of it, it has no actual surface. It has as much surface as an ocean wave, remembering that the ocean has the surface. The electron has as much surface as a light beam. The only "surface" to it is like the repulsion you'd feel between a pair of magnets. However this is quite enough to create what we perceive as a solid object. The intangible photon now appears to be a tangible mass, because it's going round and round circling at the speed of light. It's tied in one location, and whilst not idle, it's no longer going anywhere with respect to you. We could say *it's going nowhere fast*. So when you hit it, it's *you* hitting the photon instead of the photon hitting *you*. The photon had momentum, and now it's got inertia. It's got *mass*. But we don't call it a photon any more. We call it an electron.

All we had to do to create mass was "stop" the momentum of the photon. It isn't really stopped because it's still going round so now it's angular momentum. But now the photon isn't moving in aggregate with respect to you. So now it's an electron, and the travelling energy is going nowhere fast so it's "rest energy". That's what mass really is. They call it rest mass, only it's rest energy, and it's not actually at rest. When all's said and done, if it's the photon moving it's got momentum, but if it's you moving it's got inertia. That's mass. It just depends on who's moving. Because momentum and inertia are the same thing. There's a symmetry between them. Because motion is relative, and that's what relativity is all about.

We can destroy mass just as easily by bringing an electron and positron together to reverse the process. The electron and the positron are just two mirror-image trivial knots which cancel one another and annihilate, resulting in two 511 keV photons. It's like the electron is a twist in your tight taut fishing line, and the positron is the opposite twist. Slide them together and *twang*, gone. The trivial knots are no more, and the photons are off like a shot. And we do destroy mass easily. We do it every day. Annihilation is routine, we take pictures with it. It's called a PET scan, and if you're lucky there's one in a hospital near you:

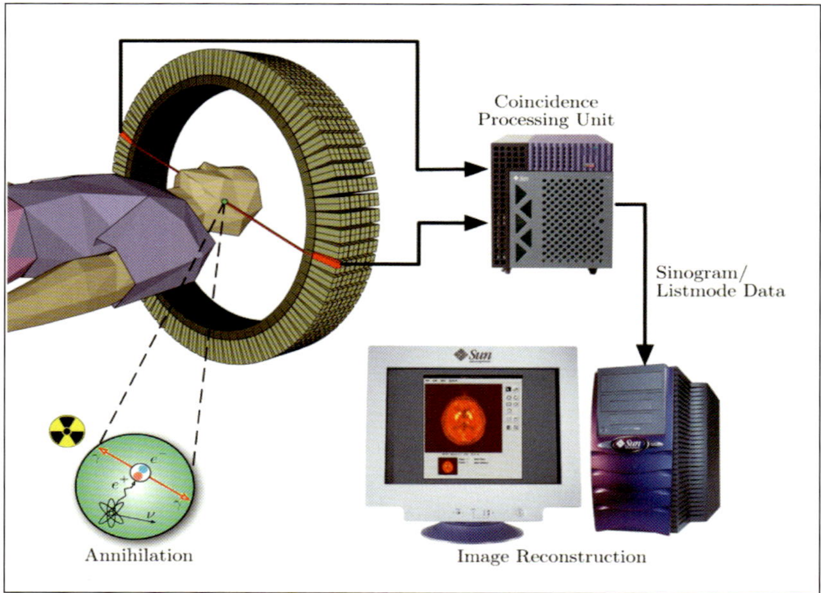

Figure 43 – PET scan

It really is that simple. Energy is usually a travelling stress like a photon, rippling through the volume of space like a shimmy shooting through a ghostly block of transparent rubber. Mass is just how you measure it when it's tied in a knot so it's going nowhere fast. Or in other words:

Mass is a measure of the amount of energy that is not moving in aggregate with respect to the observer.

GOING NOWHERE FAST

The geometry we're all familiar with is usually flat on paper, with angles and lines and the occasional curve. Geometry in three dimensions is somehow very difficult, and the only way to get to grips with it is via hands-on experience. For example a circle with a twist can be made quite simply from paper, but with a little rearrangement can look totally different. It's a strip that goes round once, and does a barrel-roll at the same time. But with a little shoving and pulling, this self-same strip can be made to look like a figure of eight:

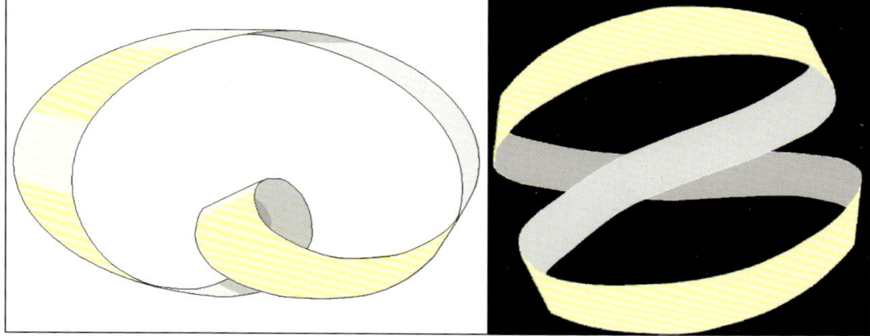

Figure 44 – Paper-strip circle with a twist, and the same strip in figure-8 form

Now the strip turns round twice, you've got *two* circles, and the barrel-roll twist seems to have disappeared. But all you have to do to get it back is pull out one loop of the eight to make it bigger. Voila, there's the twist again. Twist and turn is rather odd like that. When you push the two loops of the eight together to make one double-wrapped loop, there's no sign of the twist at all. But when you're dealing with light, you're dealing with something that has no surface and isn't flat like paper. There's a melding going on. Arranging one circle with a full twist into two circles, means that each has half a twist. Hence the möbius-doughnut electron with its dark line on page 41. Look carefully and you can see that the dark line is "the circle with a twist". The photon goes round and round because it's wrapped round twice.

Perhaps the difficulty of three-dimensional geometry is why people have so much difficulty with mass. Perhaps another reason is language. Perhaps it's a combination of both, or more, I'm not sure. But we use the word "mass" in many different ways, and it really doesn't help. The accepted definition of mass is rest mass, which is the same thing as invariant mass, intrinsic mass, and proper mass - it's defined as the total energy of a system divided by c^2.

There's also active gravitational mass which tells you how much gravity the energy causes, and passive gravitational mass, which is a measure of how much an object is attracted by gravity. People also talk of inertial mass, which tells us how much force we need to apply to accelerate or decelerate an object. It doesn't apply for a photon because it travels at c, and you can't make it go faster or slower. Then there's relativistic mass, which is just a measure of energy, which is why it applies to a photon. When you apply it to a cannonball travelling at 1000m/s, it's a measure that combines the rest mass energy with the kinetic energy into total energy.

The kinetic energy of a mass in motion is droll, but I need to revisit Compton scattering to explain why. We use Compton scattering to move an electron, and in simple terms the electron is a photon going round in a circle like this ○. The Compton scattering effects an Inverse Compton on the electron's component photon, bending it like that U-shaped steel bar I mentioned earlier. But this photon is tied in a 511 keV knot. The deflection doesn't break the knot. It stretches it like kicking a rubber ring, it alters the velocity vector and translates into motion. What you see is the same as what you'd see if you moved past the electron. It's a photon travelling in a circular path, so if it's side-on you'd see that circular path looking like a helical path. One full turn round the helix represents the relativistic mass, the total energy. The circular component of this represents the rest mass. The difference represents the kinetic energy. It tells you how fast the energy that's going nowhere fast... is going somewhere.

Figure 45 – Helix

It's ridiculous, but that's how it is. The kinetic energy tells you how fast the energy that's going nowhere fast is going somewhere. Because accelerating an electron is like trying to stretch a spring. Ever seen a split-ring spring? It's like a un-joined circle, like this: **O**. To get the electron moving we have to deform the ring into one turn round the helix. The energy required increases as we attempt to deform it further. As we accelerate more and more, the helix is effectively stretched straighter and straighter. But we can't stretch it straighter than straight. A photon travels at c, it can't travel in a straight line at c and still be going round in circles. It would have to go faster than light to do that. And light doesn't go faster than light. So matter can't travel as fast as light. No way no how, and it's easy when you know how.

We start to see something here that was always there in Einstein's simple trigonometry, and in his light beams and clocks. What we see is the chilling simple truth that underlies Einstein's postulate. You realise why we always measure the speed of light to be the same, and why we can't go faster than light. And that chilling simple truth is this: *we are made of light*. Think about it. We use pair production and annihilation to convert photons into electrons and positrons and vice versa. They're the same stuff. An electron is a "fermion". It's got mass, a measure of energy/momentum. It's a photon configuration. The photon is the mediator of the electromagnetic force. It's a "boson". It's got no mass. But all we need to do is tie it in a knot and now it does. That means we don't need to couple to the Higgs field to explain mass. We don't need the Higgs boson, because the photon is boson enough. All we're talking about is stress volumes travelling in space, and tying them into stable three-dimensional solitons.

And it gets worse. Because what this also means, is that the Large Hadron collider isn't just some search for the Higgs boson. The Large Hadron Collider is just.. *playing with knots*.

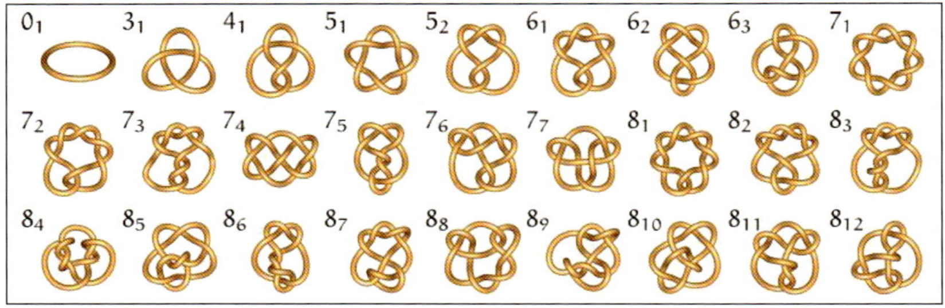

Figure 46 – Knots

CHARGE EXPLAINED

Charge is another one of those things you learn about in physics. Well, you think you do, but you don't. Not really. The textbooks don't explain it, and they shrug off this omission by telling you it's fundamental. It isn't. It's as fundamental as mass, which is not very fundamental at all. Pair production and annihilation demonstrate that charge can be created and destroyed along with mass, so it isn't fundamental. The thing is this: if you understand mass you already understand charge. But you probably don't realise it yet. So I'll explain it.

Figure 47 – Tesla coil

Let's start with the easy stuff. We know that we can rub a balloon to create an electric field. It can pick up a piece of paper or make your hair stand on end. We've all seen and felt a spark of static, blue and crackling as electricity tears the air. We know that high voltage is called high tension, and tension is negative pressure and pressure is stress. So we're happy with the fluid analogy where a current flows from the negative to the positive terminals of a battery. They got that backwards in the old days, but that doesn't matter. We just measure the rate of flow in terms of amperage, and multiply by time to get charge, and multiply again by voltage to get energy. We work out that the amount of charge in a battery is all about the number of electrons available to flow, and we know that our charged-up balloon has a surplus of them above and beyond its protons.

So, how much charge is in a flat battery? *None*, I hear you say. Wrong. It's chock *full* of charge. It's full of positive charge *and* negative charge. That's why it's got mass. That's why it's a material object. If there wasn't any charge, it would be a whole heap of gamma radiation.

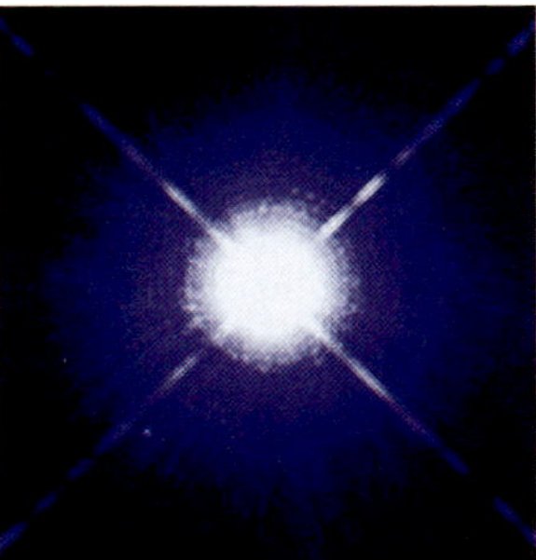

Figure 48 – Gamma radiation

But let's keep it simple and forget the positive charge for the time being. What is it about these electrons that keeps our laptops humming? What is this mysterious thing called charge? The answer is easy once you know how to see it. Go to the kitchen, get a glass, then quickly fill it with water and hold it up to the window. You'll see bubbles swirling and silvery, pop pop popping. They aren't actually silver of course, they just look that way because they distort the light. Now go to the cutlery drawer and pull out a spoon. It's silvery. Metals look that way because they're awash with electrons. When you look at a spoon you're seeing those electrons, or more properly, the charge. It's reflective, silvery. Charge looks like this for the same reason as those bubbles. It's like a highway mirage on a hot sunny day. You see what looks like water on the road ahead, but it's merely the light from the sky bent towards your eye. You are seeing distortion, and it's silvery like a bubble because it bends the light.

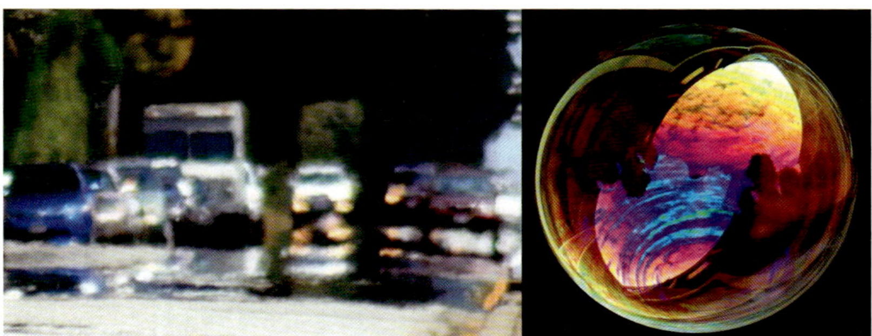

Figure 49 – Road mirage and bubble

Charge is distortion too. Charge is "curl". Charge is *twist*. If it wasn't there, your electrons would be gamma photons of 511 keV apiece. To show you how it works, I need you to play with plates. Take two dinner plates, one in each hand. Find a swimming pool or a pond, preferably on a sunny windless day. Dip one of the plates halfway into the water. Now stroke it gently forward in a paddling motion whilst lifting it clear. Notice that you create a "U-tube" double whirlpool that moves slowly forward through the water. If the pool is large enough, the double whirlpool will eventually settle down into two dimples on the surface of the water, visible as two black-spot shadows on the bottom. This is a Falaco Soliton[28], first observed in Rio in 1986:

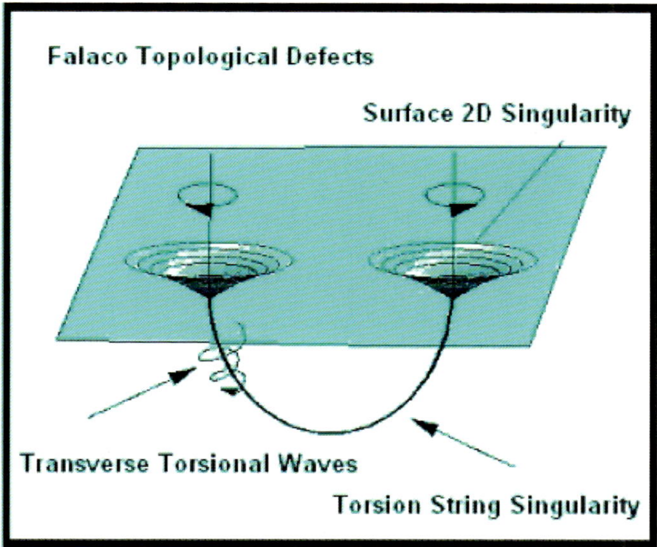

Figure 50 – Falaco soliton

It's very stable, and can persist for maybe an hour. But you don't need to wait for that. Create one double whirlpool with one dinner plate, then step to one side and create another one with the other dinner plate. You'll need a little practice, but after a while you'll have the knack of it, and you'll be able to create two double whirlpools with ease. Aim them at each other. Watch carefully. Notice what happens. When the left-hand-side of one double whirlpool is near the right-hand-side of the other, the two opposite whirlpools move together. When the left-hand-side of one double whirlpool is near the left-hand-side of the other, the two similar whirlpools move apart. What you're seeing is *attraction and repulsion*.

Now aim two double whirlpools straight at one another, face on. This is best in a shallow pond with a muddy bottom. The two double whirlpools meet and merge and are gone with a surprisingly energetic puff of muddy water. You've just seen *annihilation*.

It's another fluid analogy of course. It isn't a perfect analogy because the vacuum of space isn't a fluid like water. Space doesn't flow like some nineteenth-century aether wind. There's nothing there, but energy can travel through it, and we can talk about a photon as a stress travelling through the volume of space like a transverse wave propagating through a block of ghostly transparent rubber. Then we can understand mass by talking about pair production, where a massless gamma photon is converted into an electron and a positron. Both can be viewed as a toroidal photon configured as a möbius doughnut, a travelling stress that twists and turns to stay in place. The difference is that one twists and turns one way, and the other twists and turns the other way.

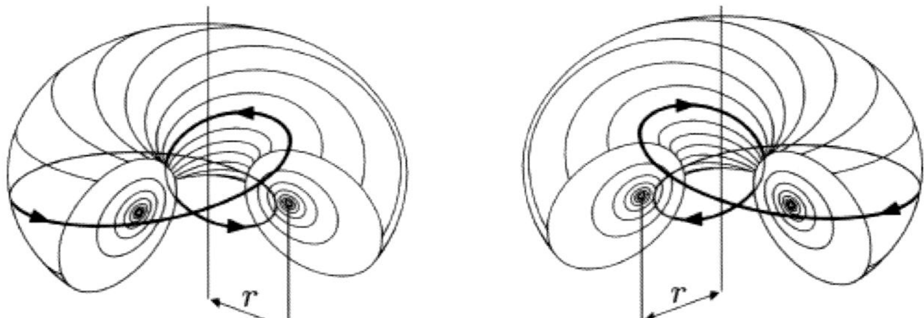

Figure 51 – Toroidal photon electron and mirror image positron

While they're travelling stresses tied as trivial knots rather than travelling fluid, the electron and positron will attract one another like the left and right whirlpools. When they meet the opposite twists cancel, and the electron and the positron annihilate to become gamma photons flying off in opposite directions like that puff of muddy water.

Of special note here is that our mental image of the electron must change. Rather than a billiard ball particle that has "got" charge extending out into space in the form of an electromagnetic field, the latter is one of the things that the electron *is*. The electron has no surface because a photon has no surface. A photon is light, and light travels in straight lines. When it doesn't, we say space itself is curved. When we consider an electron to be a photon travelling in a tight twisting loop, we must consider the space itself to be twisted. Thus charge can be construed as "twisted space", which forces a travelling stress to keep twisting and turning in one place. Since there are no solid objects to brace against in this purely geometrical concept, the only way to make a twist is to make an untwist at the same time. Hence charge is always conserved. In addition, the concept says that mass is always associated with charge. Any massive particle exhibiting zero net charge must contain hidden charge. Hence we can reason that a neutron has charge like a flat battery has charge. The positive charge is matched by the negative. Charge is present, but is not apparent while the neutron is within a nucleus. However should the neutron escape that nucleus, in circa fifteen minutes it decays into an antineutrino, an electron, and a proton, and the charge is then revealed. It's a twist in the thing you call space, stretching out into space. This is why you could call an electric field a "twist field". Let's see how it affects an electron.

In simple terms an electron can be viewed as a photon travelling in a circle and looking like this: ○. Drop the electron into a cube of space so it looks like this: ○ . If we take a side view of our component photon at one instant in time, we can imagine a vertical slice through the möbius doughnut, like this: | . When we twist the cube from top to bottom, the vertical slice is tilted like this: / . So the photon will travel downwards while it's also travelling in a twisting turning loop. Hence the electron digs down through the electric field like a drill bit. That's how attraction works. Repulsion is the same sort of thing, but of course a positron goes the other way, like a drill in reverse.

Note that the electric field isn't just a twist in one dimension, it's actually in three dimensions. Your electron digs down like a drill bit from any direction. But it's very difficult to think in three dimensions. Our primary input is visual, and whilst binocular vision permits depth perception, we tend to think in two dimensions. That's why getting the *feel* for something is what intuition and grasp are all about. It gives us a better, three-dimensional concept. To appreciate this, get a lump of Plasticine or maybe the wax from a Babybel cheese, and make a cube. Now try twisting it in three dimensions. Two twists is easy: twist, turn, twist. But doing the third one is surprisingly difficult. In the end you have to just do it by feel: twist turn, twist turn, twist. You end up with something like this:

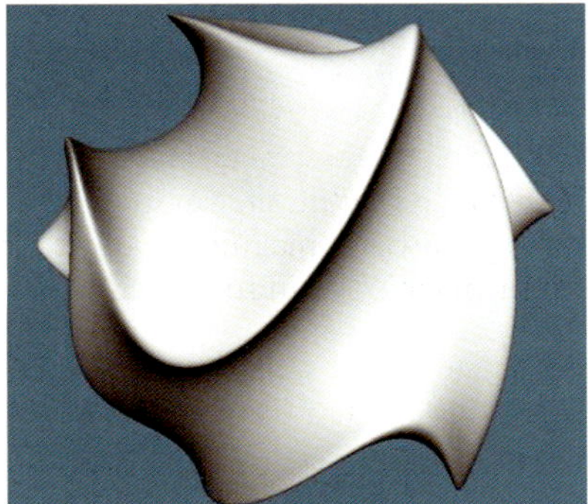

Figure 52 – Twisted cube

The easiest way to get your head round the geometry is to imagine that the twisted cube is a twisted block of water, and we've got to swim through it. As you're swimming behind me you find that all the twisting and turning means you've got to swim further than you thought, and you come out of the other side gasping for air. But you can now understand refraction. Light travels slower through a glass block because it's got to make its way through all that twisting and turning in all directions, be it positive or negative.

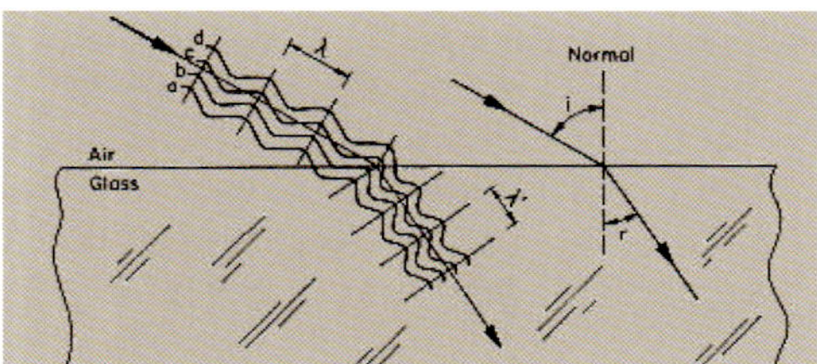

Figure 53 – Refraction

Talking of turning, let's talk about magnetism. Imagine that you're flying through space, but the space ahead of you is twisted like a Catherine wheel or pinwheel because of the electric field.

Figure 54 – Pinwheel

Hold your arms out and walk forwards like you're an aeroplane. When you encounter the twisted space, your wings will tilt. The twisted space will make you rotate. It will make you turn. We now use relativity to work out that if you aren't travelling through twisted space but you find yourself turning, then the twist must be travelling through you. That's what happens when a current flows through a wire. Imagine the current is flowing down a wire from your eyes into the page. This introduces an anticlockwise twist:

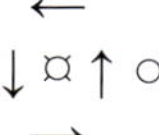

The nearby electron ○ is basically a circling photon. This comes full circle in the twisting space before it's gone round 360 degrees. So it ends up at a different place, and describes a cycloid motion. It really is that simple. The electric field is effectively a "twist field", and if you move through it you perceive a magnetic field, which is effectively a "turn field". It's so obvious once you see it. And you *can* see it. You can see how a magnetic field changes the polarization of a beam of light via the Faraday effect:

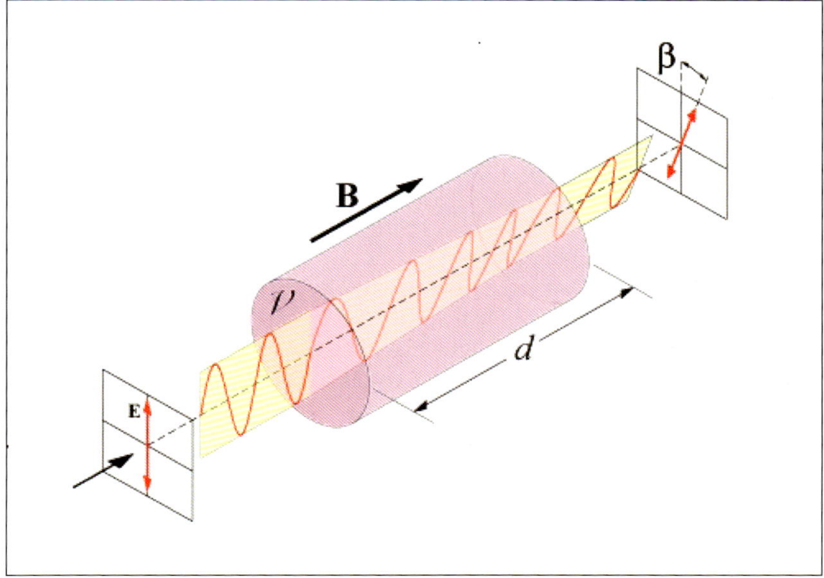

Figure 55 – Faraday effect

That's the utter simplicity of electromagnetism: twist and turn. It tells you a battery is like a wind-up clockwork spring, only the twist is in space rather than steel. The electric twist extends forward with the flowing current, and it makes things turn like a pump-action screwdriver. This is the principle of the electric motor. But you can turn a screw with an ordinary screwdriver too, extending the twist forward. That's the principle of the dynamo. Applying a forward motion to the twist achieves a turning motion, and vice versa. It's even better if you feel it by playing with a drill bit, pushing it through your fingers and feeling them turn, or turning it and feeling the push. It's beautiful, it's simple, and it works for every electromagnetic phenomena there is.

Most materials aren't magnetic because all this twisting and turning is equal and opposite in all directions, even for your charged-up balloon. It's what you call isotropic. When it isn't, that's when you get a magnet. Fly through an electric field or past a stationary electron, and you experience more twist in the direction of travel, so you "see" a magnetic field that makes you turn. Move an electron towards you and you get the same effect. And all you need to do to make an actual magnet is arrange the atoms so that the electrons jitter round in the same orientation:

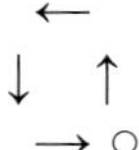

The electron is moving in a circular fashion, so its component photon doesn't need to complete a full 360 degrees to turn around. It's like the Earth going round the Sun: a day is less than one full rotation of the Earth, lasting 23 hours 56 minutes instead of 24 hours. As far as the electron is concerned there's a component of the "turn" left over, and you end up with a magnetic field similar to what you'd see if you flew past a stationary electron. It's rather like the inverse of the current in the wire situation, but with no current and no wire.

Whilst I describe a magnetic field as a "turn field", you have to remember that space is like a ghostly elastic solid with stresses rippling through it. The electric field is the "twisted space", and the magnetic field is only your relativistic view of this when you move through it. There are no actual regions of space turning round and round like roller bearings. That's why you can't have magnetic monopoles. But you *can* have superconductors. High temperature superconductors currently consist of copper oxide planes.

The atoms present an array of opposite magnetic fields rather like a conveyor belt, allowing electrons to zip through effortlessly like they're not moving at all.

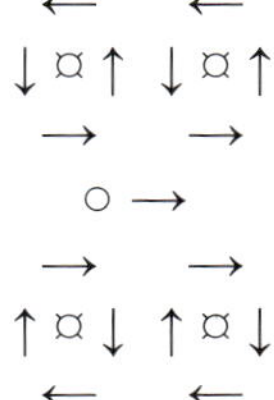

The array of magnetic fields act like *wheels*. But of course, it's a little more involved than that, because wheels need bearings and axles. Here's a picture of a high-temperature superconductor called yttrium barium copper oxide, or YBCO for short. The chemical formula is $YBa_2Cu_3O_7$ and it's a crystal so you get repeating groups. In simple terms the "wheels" are where the pyramids are:

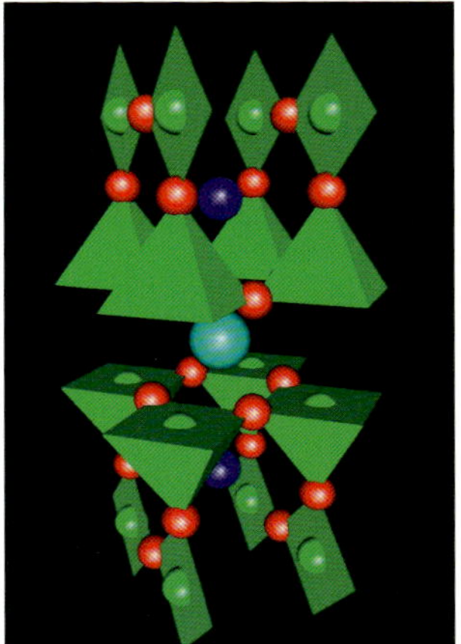

Figure 56 – YBCO superconductor

Low temperature superconductors aren't quite the same. You have to think *Barn Dance*, where you're an electron with a "Cooper pair" dance partner making your own magnetic fields as you go. When everybody's cool, the dance line is tidy and you swing easily from one end to the other. But when it's hot and late and people are bumping around all tired and emotional, you spill somebody's beer, lose your partner to a "phase slip", and get into a fight. Interestingly, in both types of superconduction the superconductor is what's called diamagnetic. It doesn't get magnetised because of the Meissner Effect, where all the internal opposite magnetic fields scramble an applied magnetic field like a billion egg whisks, hence it doesn't get very far into the material.

Figure 57 – Meissner effect: levitation of a magnet above a superconductor

All interesting stuff. And so very simple. The twist and the turn are just two aspects of the same thing. That's why we have electromagnetism and the electromagnetic field. An electric field is the same thing as a magnetic field, it just depends how you're looking at it, and whether you've got relative motion. That's relativity for you. Once you learn how to see things the way they are, things get a whole lot simpler:

Charge is curl, charge is twist. The electromagnetic field is a region of twisted space, and if we move through it we perceive a turning action which we then identify as a magnetic field.

It's obvious once you see it. A whirlpool isn't a particle that's "got" rotation. Rotation is part of what it is. An electron isn't a particle that's "got" charge. Charge is part of what it is. Charge can be created and destroyed in pair production and annihilation along with mass, so it isn't fundamental. Instead it's the product of a particular geometric configuration of space. And this geometrical concept of electromagnetism that talks of twisted or "curved" space forces us to re-examine our view of curved spacetime. But to do that, we have to examine our very view itself.

REFERENCE FRAMES

When you move through an electric field you see it as a magnetic field, and when you move through a magnetic field you see it as an electric field. Because it's the same old thing really, it's an electromagnetic field, and the difference is down to you. Sometimes you don't realise that things are the same because you see them a particular way. Because you walk around all your life wearing some very special sunglasses. They're like Ray-Bans. You grow up with them, so much so that you don't know you're wearing them. They colour your vision but you can't see how. They stop you seeing the way it is. But these things that colour your vision aren't sunglasses. What they are is *reference frames*.

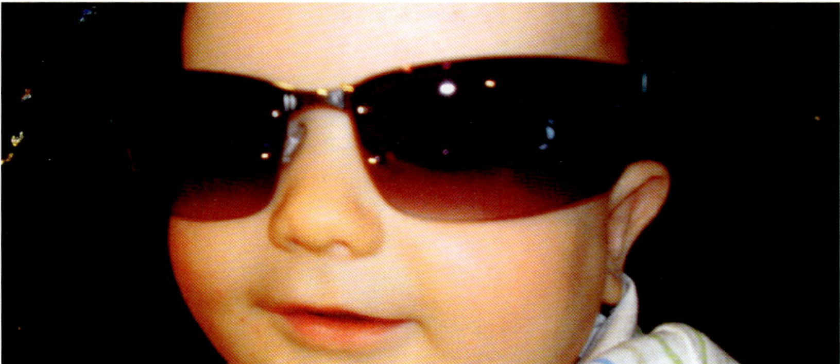

Figure 58 – Reference frames colour your vision

Let's have another thought experiment. It involves my spaceship again. You're the new co-pilot, and I've just taken you down to the cargo bay and shown you *the box*. It's a ten foot canister sitting in the airlock. It's comfortable, cushioned, equipped with an air supply and thrusters. Actually it's an old escape pod, but we call it the box because it doesn't have any windows. And as you may have gathered, you have to get in the box.

You climb in, find the light switch and click it on, then I swing the hatch closed and give you a *slamslam* goodbye. You hear me leave then the pumps evacuating the air, followed by the grinding of steel as the outer doors open. There's a jolt as the launch ram pushes you outside the ship, and then everything goes quiet. There are no windows in this box. There is however a radio so we can stay in touch. I call you up to say I'm pulling back a little, leaving you there in your box.

Figure 59 – Zero gravity

You say *A-OK catch you later*, and enjoy a few zero-gravity games. You do some somersaults and push yourself from one side of the box to the other playing racing turns. After a while you find the tennis ball in your jumpsuit pocket and throw it against the opposite wall, smiling when it bounces straight back into your waiting hand. The radio bleeps into life and it's me asking how you're doing. *Fine* you reply, I ask if you can detect any acceleration and you say *Nope*. You're cool, because you're weightless, it's fun. You're in the box. There are no windows. You pitch the ball across the inside of the box and it goes straight as a die, bounces off the side, and back into your hand. You know you're in an "inertial reference frame". You can feel no gravitational force, and you can detect no outside forces acting upon the ball. You are not accelerating.

I ask if you'd like to try a little acceleration, you say *Sure*, and then you feel and hear the thrusters burning. You find yourself pressed back, and now one side of the box feels like it's the floor. You stand up and it's like being on the surface of the moon, or shipside under artificial gravity. You can feel your weight. You throw the ball and it travels in a lazy arc towards the floor and bounces around a little before settling there. You're now in a non-inertial reference frame, and you know you're accelerating. You can feel it, you can measure it locally, within your frame, within your box.

I shut off the thrusters so you're back weightless again, floating in your new inertial reference frame. You can no longer feel any acceleration. You can't measure any. The ball flies straight as a die. As far as you're concerned, you're not accelerating.

Did I mention that you're falling into a black hole? No? Sorry about that.

Here's how it is: a body in freefall is not accelerating. If you say it is, you're mixing reference frames. You're looking at yourself from some reference frame out somewhere in space rather than from *your* reference frame, that of the body in freefall. Don't be mistaken about this. The Principle of Equivalence between gravity and an accelerating box applies when you're standing on a planet, *not* when you're in freefall.

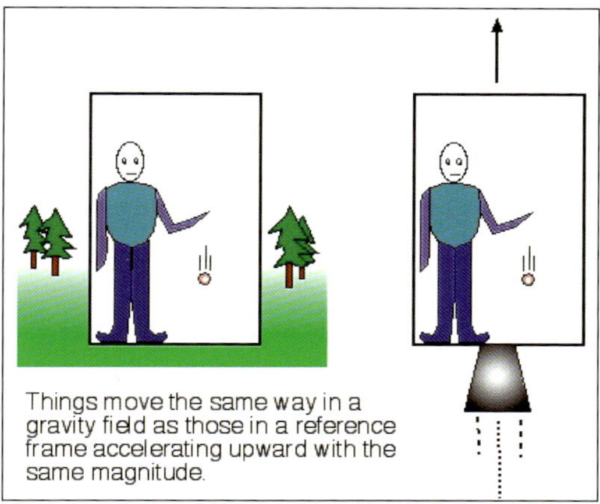

Figure 60 – Principle of equivalence

There really *is* no force acting on your freefalling body. You can't feel any, you can't see any, you can't measure any. Because there isn't any. If you looked out of a window to see a moon going backwards, you're mixing frames. This is why the force of gravity is not technically a force. It exerts no force on you. In your reference frame you are not accelerating. You never can be, because in *your* reference frame *your* velocity must *always* be zero. If you say it isn't, you're mixing frames again. You really are not accelerating, and that's why you can throw that ball straight. You don't accelerate when you fall down, but you do accelerate when you don't. And that's why Einstein called gravity a pseudo force.

Meanwhile I'm sitting out in space in the ship where the black hole gravity is so negligible and swamped that I can consider myself to be in an inertial reference frame too. All I have to do is switch off the ship gravity, and I can play ball games just like you. But when I look through the viewscreen I see you falling faster and faster towards the black hole, knowing that as far as you're concerned, you're in an inertial reference frame just like me. We're both in inertial reference frames, but yours isn't inertial as far as I'm concerned, and nor is mine to you. What's happening is that your inertial reference frame is changing and you can't see it.

But you can detect *something*, if you have the right equipment. And that you do. The radio crackles and it's me again, telling you to open a concealed hatch and pull out a scientific instrument that looks like a long thin dumbbell. I tell you to perform a *Pound-Rebka* experiment, and you follow my instructions and find that you get a photon blue-shift reading when the instrument is pointed in a given direction. I tell you it's pointed at the black hole, and what you're measuring is a slight tidal force in that direction.

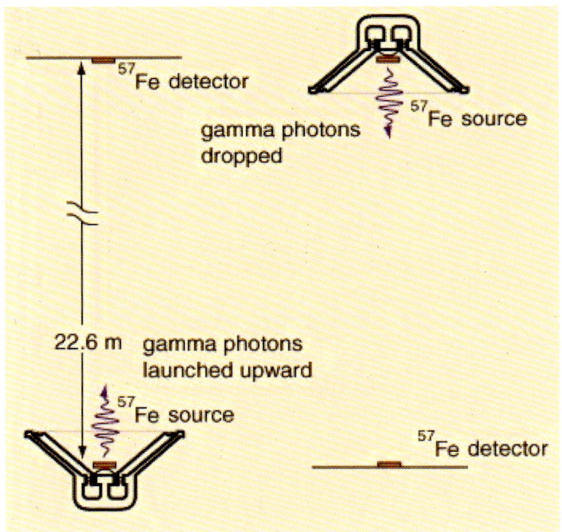

Figure 61 – Pound-Rebka experiment

Note that a "proper" gravitational field is never uniform. There's always a tidal force. Sometimes the tidal force is "negligible", but don't neglect it because it's never ever zero. If gravitational fields were uniform, they'd be hills without slopes, they'd be flat, so they wouldn't be hills at all, and things wouldn't fall down them. The tidal force is *always* there because of a slight change in the gravity in a given direction. This is where the Principle of Equivalence isn't quite perfect, because there is no tidal force in a simple accelerating box in free space. Einstein knew this, but people only look at what he generalised in 1911 and take it all too far. Here's a quote from a physicist called John R Ray dating from 1977:

"The first thing to note about the 1911 version of the principle of equivalence is that what in 1911 is called a uniform gravitational field ends up in general relativity not to be a gravitational field at all – The Riemann tensor is here identically zero. Real gravitational fields are not uniform since they must fall off as one recedes from gravitating matter".

It's really obvious when you think about it. But people never do. They never actually look *at* the reference frames that they look *through*. You're in your box, and gravity is continuously changing your inertial reference frame. But you're *in it*, you're *immersed* in your reference frame, you can't see it. It's what you look through to see the world, you can't see it changing. All your photons are blue-shifting, but you don't notice it. Your time is dilating, but you don't notice it. Your seconds are changing, and your metres too. But not to you, because they're part of what you are. Your reference frame is how you see the world, but it isn't how the world is. Because like I said, in *your* reference frame, *your* velocity is always zero, and it puts you at odds with Copernicus. Because in *your* reference frame, you don't move, so *the Sun goes round the Earth*. That puts you into crystal sphere territory, and you need to break out:

Figure 62 – Crystal spheres

I'm not that fond of reference frames. They aren't real, they're an abstraction, and they get in the way instead of making things clear. Let's see them for what they are, and learn to look at the world the way it really is. So brace yourself now, full reverse thrust, come on home. We haven't got time to let you fall into that black hole. Because to understand gravity you have to understand time. Aristotle, born 2392 years ago and counting, said *Time is Suspect.* How right he was. Time is the key. Understand time and you understand relativity like never before.

TIME EXPLAINED

Time is very simple, once you get it. But "getting it" is so very difficult. That's because your current concept of time is so deeply ingrained. It's been that way since you were a child, when you formed a mental map of the world using your senses and your brain and the things you were taught. You use this mental map to think, and you are so immersed in it that you can't see things the way they really are. You can't see that the map is not the territory. You can't see that you're locked into an irrational conviction that clocks *run*, that days *pass*, that time *flows*, and that a journey takes a *length* of time. You have an ironclad mental picture of time as a dimension of length, offering freedom of motion like the dimensions of space. And that picture is wrong.

It takes steely logic to break out of this conditioning. It takes cold rational thought, and you need to examine something you take *so* much for granted that you never *ever* think about it. First of all we need to look at your senses and the things you experience. Let's start with sight. Look at the picture below:

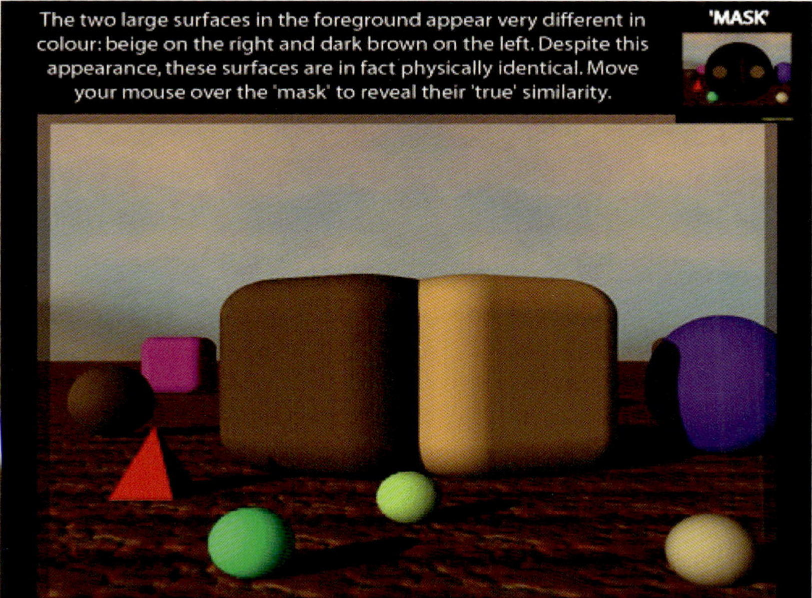

Figure 63 – Brightness illusion by R Beau Lotto

This is another optical illusion. Apart from the shading where they meet, the two large rounded blocks are actually the same colour. Fold the paper a little to compare them. Then you can see that the difference is merely an illusion. What this tells you is that colour is subjective. It isn't a real objective property of things in the world. It's a perception, it's in your head. Colour is a "quale", one of the "qualia", the way things seem. Light doesn't actually have a colour. It has an energy, an oscillation, a frequency. What it's got is a *motion*.

Let's move on to sound. Imagine a super-evolved alien bat with a large number of ears, like a fly's eye. This bat would "see" using sound, and if it was sufficiently advanced it might even see in colour. But we know that sound is pressure waves, and when we look beyond this at the air molecules, we know that sound relies on *motion*.

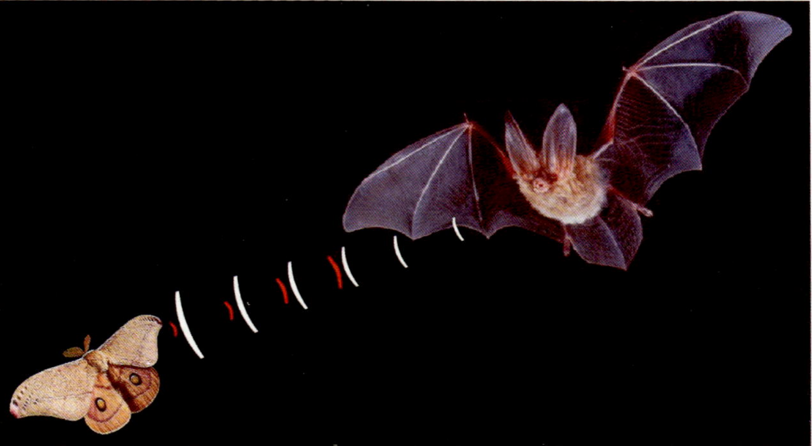

Figure 64 – Bat sonar

Pressure is related to sound, and to touch. You feel it in your ears on a plane, or on your chest if you dive. This pressure of air or water is not some property of the sub-atomic world. It's a derived effect, and the Kinetic Theory of Gases[29], tells us that pressure is derived from *motion*.

You can also feel kinetic energy. If a cannonball in space travelling at 1000m/s impacted your chest you would feel it for sure. But apologies, my mistake. It isn't the cannonball doing 1000m/s. It's you. So where's the kinetic energy now? Can you feel it coursing through your veins? No. Because what's really there is mass, which is energy, and *motion*.

You can also feel heat. Touch that stove and you feel that heat. We talk about heat exchangers and heat flow as if there's some magical mysterious fluid in there. And yet we know there isn't. We know that heat is another derived effect of atomic and molecular *motion*.

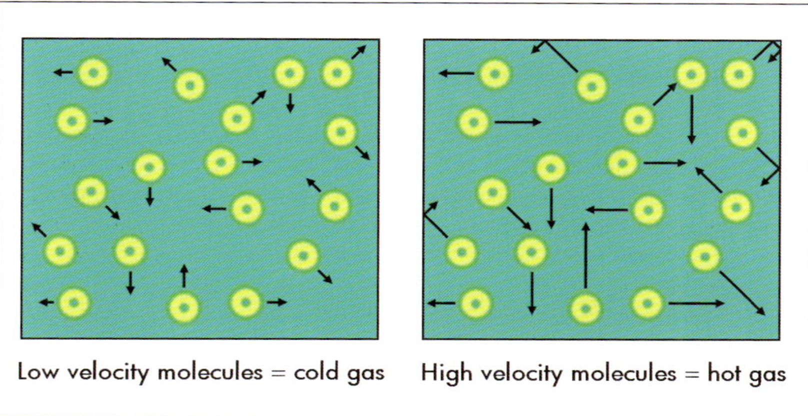

Figure 65 – Heat as molecular motion

Taste is chemical in nature, and somewhat primitive. Most of your sense of taste is in fact your sense of smell. Do you know how smell works? Look up olfaction and you'll learn about molecular shape. But a theory from a man called Luca Turin[30] says it's all down to molecular vibration, because isomers smell the same. That's *motion* again.

The point of all this is there's a lot of motion out there, and most of your senses are motion detectors. But it probably never occurred to you because you're accustomed to thinking about the world in terms of how you experience it, rather than the scientific, empirical, ontological things that are *there*. And nowhere is this more so than with time.

So, what is time? Let's start by looking at the definition of a second as given by the International Organization for Standardization under ISO 31-1:

Under the International System of Units, the second is currently defined as the duration of 9,192,631,770 periods of the radiation corresponding to the transition between the two hyperfine levels of the ground state of the caesium-133 atom....

A duration is a period, so a second is nine billion periods of radiation. But what's a period? We know that the radiation here is the thing we commonly call light. We also know that light has a frequency. So let's look at frequency and try to tackle it with mathematics:

$$\text{Frequency} = 1 / T \quad \text{and} \quad \text{Frequency} = v / \lambda$$

This says frequency is the reciprocal of the period T, and is also velocity v divided by wavelength lambda. Combining the two, we can say $T = \lambda / v$, which means a period T is a wavelength divided by a velocity. We know that a wavelength is a distance, and we know that a velocity, or more properly a speed, is a distance divided by a time. So if a period is a wavelength divided by a velocity, that means a period is a distance divided by a distance divided by a time. We can combine $T = \lambda / v$ and $v = \lambda / t$ and write it down as:

$$T = \lambda / (\lambda / t)$$

Then we can cancel out the λs to get $T = 1 / (1 / t)$ then cancel the double reciprocal to leave $T = t$. The answer to the question "What is a period?" is $T = t$, which only tells us that a period of time is a period of time. This demonstrates how mathematics cannot be usefully employed on its own axiomatic terms, and how the official definition of the second is circular. It tells us nothing about the true nature of time. What *is* its true nature? How do we get to the bottom of it? Let's look at frequency some more:

Frequency is the measurement of the number of times that a repeated event occurs per unit of time.

Our unit of time is the second. Frequency is the number of events per second. A second is nine billion periods of electromagnetic radiation. A period of radiation is an electromagnetic event, caused by an electromagnetic event happening inside an atom. The event concerned is the hyperfine transition. It's associated with magnetic dipole movement. It's a *movement*. For an event to happen, *something has to move*. Some component of the caesium atom has to travel some distance. The magnetic dipole movement is a spin-flip interaction between the nucleus and an electron. It's magnetic, so it's electromagnetic in nature. Like the electron is electromagnetic in nature. Like the photon is electromagnetic in nature, because the photon is the "mediator" of the electromagnetic force. So in some simple respect, we can consider some vital component of the atom to be electromagnetic just like light.

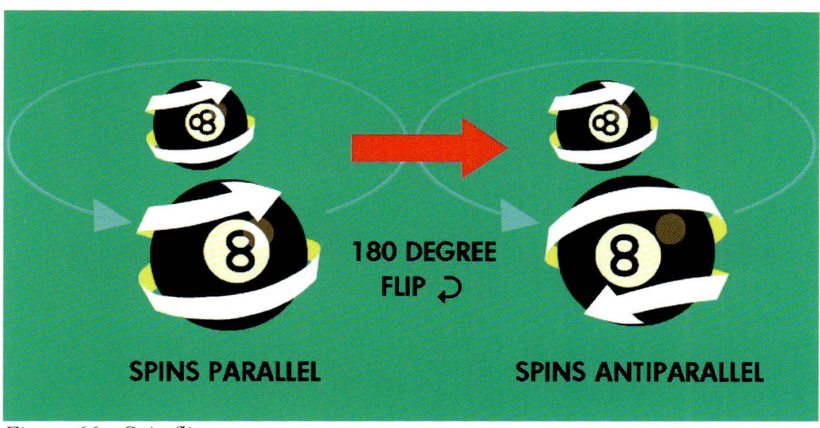

Figure 66 – Spin flip

We recall pair production and annihilation, and the answer comes with a rush: it's some kind of *light* moving inside the atom. It causes more light, radiation, photons with a frequency, a motion, and we count nine billion motions and call it a second. Then we use this second to measure the speed of light. In caesium atoms, in hydrogen atoms, in our own atoms, in the atoms of everything. And we measure it using metres:

The metre is the length of the path travelled by light in vacuum during a time interval of 1/299792458 of a second...

No wonder the measured speed of light never changes. The second is defined by the motion of light, and so is the metre. That means we measure the speed of light *only in terms of the speed of light*. And so the penny drops: the mathematics is circular because time is circular. The interval between two events is measured in terms of other events. And the interval between those events is measured in terms of other events. Until there are no events left, only intervals. And intervals are frozen timeless moments. In a universe that is totally frozen with no motion and no events, including events within the mechanism of the observer, the concept of time cannot apply. When you stop the clock you stop *motion*, not time. You need events, not frozen timeless intervals to mark out the time. The events aren't in the time, the time is in the events. Because time is merely the measure of events, or change, or motion, measured against some other events, or change, or motion. The conclusion is stark, and it is undeniable: *you don't need time to have motion. You need motion to have time.*

You don't need regular atomic motion to mark out time. Any regular motion will do. Yes, we counted nine billion oscillations and called it a second. One, two, three… nine billion. But you don't have to count hyperfine radiation from a caesium atom. You could count beans in a bucket. *Ping, ping, ping,* chuck them in, regular as clockwork.

Figure 67 – Beans in a bucket

You're sitting there counting beans into the bucket, *ping, ping, ping,* regular as clockwork. Now, what's the direction of time? The only direction that's *there*, is the direction of the beans you're throwing. A fuller bucket is not the direction of time. *More beans* is not the direction of time. The direction of time is the direction of your *counting*, and I could have asked you to count the beans *out* of the bucket. There is no real direction. It's as imaginary as the direction you take when you count along the set of integers:

1 2 3 4 5 6 7 8 9 10 11 12 13 14 15 16 17 18 19 20 21 22 23 24 25 26 27 →

It's imaginary, so you can't actually point in this direction. Nor can an arrow. There is no *Arrow of Beans,* so there is no *Arrow of Time.* And since there's no direction of time, there's no direction of time you can possibly travel in. And since you can't travel in time, you can't travel a length of time, and a length of time can't pass you by. It's all abstraction, a false concept rooted in the language we use to think. Yet we never ever think about what the words actually mean. Instead we say *the clock is running* as if a clock is an athlete, knowing full well that clocks don't really run. We say *the day went quickly* but it didn't go anywhere. We say *years pass*, but whoosh, they don't go by like buses. The *only* directions that are there, are the directions of the *spatial* motions that make the events that we use to measure the intervals between the other events. What's there is the motion of light, the motion of atoms, and the motion of clocks and buses. What's there is the motion of the Earth, and the Sun, Moon, and stars. All these motions are being counted, incremented, added up. We count regular atomic motion to use as a ratio against some other motion, be it of light, or clocks, or buses. All of these things have motion, both internal motion and travelling motion. And all those motions are real, with real directions in space. But the time direction isn't real. It's as imaginary as a trip to nine billion.

Figure 68 – Clocks don't really run

That's why the past is only in your head, in your memory, and in your records. It isn't a place you can travel to. It's just the places where things were, and all those places are still right here in the universe, right here, right now. And while the past is the sum of all nows, now lasts for no time at all. Because there's no time like the present, and time needs events, and when you take away the events, you take away the time. A second isn't some slice of spacetime. It's just nine billion motions of the light from a caesium atom. Accelerate to half the speed of light and a second is still nine billion motions of the light from a caesium atom. But there's not the local motion there used to be, because of the motion through the universe.

That's the thing we're interested in. The universe. That's the thing that's out there, the thing we're a part of, the thing we're trying to understand. When we look up to the night sky we don't see world lines, or light cones. Nor do we see a block universe. Instead we see space, and motion through it. If we consider a cube of space containing molecules of gas, we can measure the height width and depth, the three familiar dimensions of space. We have freedom of motion in those dimensions. When we measure this motion, we then derive a fourth dimension, the dimension of time, and commonly imagine that the molecules are moving *through* this dimension. They're not.

The molecules are moving *through space*. They're *only* moving through space, not through a time dimension.

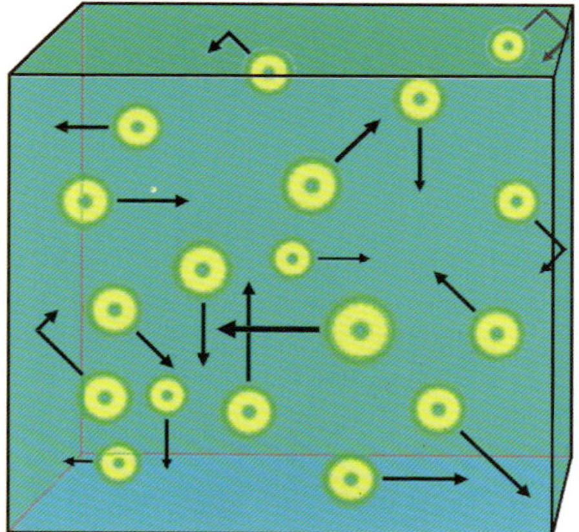

Figure 69 – Molecules

Look at the picture. What can you see? What can you measure? You can measure the height. You can measure the width. And if it wasn't just a picture you could measure the depth. That's three Dimensions, with a capital D because we have freedom of movement in those dimensions. What else could you see? What else could you measure? You would see the motion. You could measure it. You might *imagine* a fourth dimension, a time dimension. But it simply isn't there. What is there, is space, and motion through it. The motion is through space, not through time.

We can examine the notion of motion through time by considering temperature. Temperature is an aspect of heat, an emergent property, a derived effect of atomic and molecular motion. When we measure a temperature, we're measuring an average motion. Whilst we commonly speak of a "higher" temperature, we're fully aware that we cannot literally climb to a higher temperature. Hence we don't consider temperature to be a dimension. Historically however, temperature *was* considered to be a dimension, but the word dimension has gradually shifted from its original meaning of "measure", and is now assumed to offer a degree of freedom, such that we can "move" through it. Time is like heat in that it too is an emergent property or derived effect of motion, but it isn't an average motion, it's a cumulative measure of motion used in the relative measure of motion compared to the motion of light. And just as we can't travel in temperature because there's no real height to a high temperature, we can't travel in time because there's no real length to a long time.

Time offers no degree of freedom. This is patently evident: I can hop backwards a metre but not backwards a second. It's further evident when we consider speed, which is a rate of change of place. The change of place involves motion by definition, and travel through space most definitely occurs. But there can be no sense of travelling through the factor that tells us the rate of this motion through space. And since speed is defined using distance and time, and time is defined by 9,192,631,770 motions of light, in truth there are no absolute units of time. The units of time are thereby rendered subjective and relative rather than absolute and objective, and that's what relativity was always all about.

Special relativity tells us that your relative velocity alters your measurement of space and time compared to everybody else. You increase your velocity, and space appears to contract by a factor of $\sqrt{(1-v^2/c^2)}$ while time dilates by a factor of $1/\sqrt{(1-v^2/c^2)}$. If you travel at .99c, space appears to contract to a seventh of its former size, and your trip to a star seven light years away only takes you a year. But physics is about the universe, and in that universe it took seven years. The star didn't flatten into a disc because you flashed by. The space in the universe didn't really contract because *you* travelled through it. But your time is subjective, so *your time did*. And behind it is just Pythagoras' Theorem applied to the motion of light between parallel mirrors. We're working out the height of a triangle where the base is your velocity v as a fraction of the speed of light c, and the hypotenuse is 1 because that's the light path. It's just $\sqrt{(1-.99^2)} = \sqrt{(1-.98)} = \sqrt{(.02)} = .1414$, which equals a seventh.

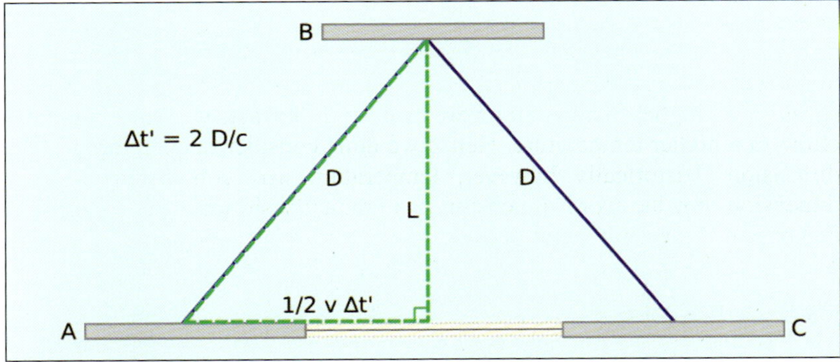

Figure 70 – Time dilation is based on Pythagoras' Theorem

Imagine yourself as a metronome, where the wand ticks back and forth at the speed of light. Each tick is a thought in your head, a beat in your heart, a second of your time. If you're motionless with respect to me I see you ticking like this: ||| . If you flash by in your spaceship, I see you ticking like this: /\/\/\. Travel even faster and you tick like this: /\/\/\. If you could travel at the speed of light, your ticks would flatline like this ―――― because any transverse motion would cause c to be exceeded. You wouldn't tick for me, you wouldn't tick for you, and you wouldn't tick for anybody else in the universe. For you, time wouldn't "pass", because you can't experience any local motion.

It isn't mysterious, and once you appreciate the history and understand where Einstein was going, you understand the utter simplicity behind it. The speed of light was always the problem, time was always the answer, and time is very simple once you get it. Here's how it works. This is the true meaning of relativity:

Time exists like heat exists. It's real because it does things to us. But just like heat it's a derived effect of motion. If the universe consisted of two objects and two objects alone, separated by some distance, we could hold a concept of space. But if those objects don't move, we can't hold a concept of time. When those objects do move, then and only then can we conceive of time, for it's not space and time that are cofounded, it's motion and time that are cofounded.

We observe three dimensions of space, plus motion through it. We measure this motion with distance and time. But our distance and time is defined by the motion of light. Therefore we only measure the speed of light in terms of the speed of light, so c is a conversion factor between our measure of distance and our other measure of motion, the measure we call time. Motion is king, because that's what we see. And the speed of light is the ultimate motion, how fast things happen, the inescapable property of photons and those electromagnetic things from which we're made.

We don't travel in time at one second per second. We travel in space, and we don't travel in time at all. The internal workings of a clock is called a *movement*, and clocks clock up *motion*, not time. Relativistic clocks don't travel through time at different rates, they travel through space at fractions of c. If they collide, they collide at the same location at the same "time", whatever their faces say is *local time*. To travel backwards in time we'd need to unevent events, and we'd need negative motion. But motion is motion whichever way it goes. You can't have negative motion, just as you can't have negative carpets. So we can't travel back in time. There *are* no time travel paradoxes, because there *is* no time travel, and there is no time travel because time is merely a measure of motion, and motion is travel, and you can't travel through travel.

So time isn't absolute, it isn't fundamental. It isn't a dimension like the dimensions of space. It's subjective, an imaginary component of abstract Minkowski spacetime. But the map is not the territory. The mathematical space is not the space. We don't see a block universe, we see a universe in motion. We don't see four dimensions, we see 3+1. The world lines are only in your head. There's no place that's the future, and there's no place that's the past. There's only the present, where the time is always now. There was no "beginning of time", because time didn't start in the first place. It was *motion* that started *in the first place*. It was a place, not a time. And it's *this* place, the place we call the universe, marked out by every light path you can track through timeless space.

Figure 71 – Clocks clock up motion, not time

We've come a long long way, in no time at all. And now we can move on. Because now we really understand what Einstein's light clocks and his electrodynamics were really all about. Now we really understand why the speed of light always looks the same: *we are made of light*. It means spacetime is a mathematical space. It's an abstraction. What's out there is actual space, and motion through it. World lines are the wrong concept, and light cones, and the block universe. They've been leading us astray when the simple answer has been overlooked:

Time exists like heat exists, being an emergent property of motion. It's a cumulative measure of motion used in the relative measure of motion compared to the motion of light, and the only motion is through space. So time has no length, time doesn't flow and we don't travel through it.

GRAVITY EXPLAINED

You probably think of gravity as "curved spacetime". Surprisingly Einstein didn't. *The Foundation of General Relativity* published in 1916 doesn't use the phrase. So neither should you. To understand gravity you have to take the ontological view. You have to learn to see what it *is*, not what it *does*. You have to learn to see the cause, not the effect. And to do that, you have to put time to one side, because time isn't the same kind of dimension as the dimensions of space. Because spacetime is space with motion through it. Yes, an object passing a planet traces a curved path, but you don't stare up at a plane and decide it's a silver streak in the sky. You take a mental snapshot, *flash*, a picture of it *now*, a timeless instant. It's the same with gravity. Take the derivative of that curved spacetime. What you get is a gradient. And it's a gradient in *space*, not curved spacetime.

But let's tackle it an easier way, via an old favourite. Think about a cannonball sitting on a rubber sheet. The cannonball is heavy, and it makes a depression that will deflect a rolling marble, or even cause the marble to circle like an orbit. It's a nice analogy, but it's wrong. It's wrong because it relies on gravity to pull the cannonball down in the first place. It's circular. It uses gravity to give you a picture of gravity.

Figure 72 – Rubber sheet analogy for gravity

To get a better handle on it, imagine you're standing underneath the rubber sheet. Let's make that a silicone rubber sheet. Silicone rubber is transparent, like my snorkel and mask. Grab hold of the rubber around the cannonball and pull it down further to give yourself some leeway. Now transfer your grip to the transparent silicone rubber itself. Gather it, pull it down some more. Now tie a knot in it underneath the cannonball, like you'd tie a knot in the neck of a balloon. Now get hold of the knot, pull it *all* the way down, and let go. *Boinggg!* The cannonball is gone. Forget it.

Now, what have we got? We've got a flat rubber sheet with a knot in it. The knot is a stress configuration that keeps the rubber all tied up so it stays in place. There's a lot more rubber where the knot is, and the knot represents energy. Or matter if you prefer. A mass. Surrounding the small central region of stress is a much larger region of tension extending outwards in all directions. Whenever you have stress you have tension too. It isn't always obvious, but it's always there, like action and reaction. Here in our rubber sheet the tension gradually reduces as you move away from the knot. So if you could measure it, you would measure a radial gradient.

Figure 73 – Knot in a sheet

But measuring it is trickier than you think. Because in this analogy we can't use a marble rolling across a rubber sheet. This rubber sheet represents the world, there's no stepping outside of it. Our "marble" has to be *within* the sheet, and a part of it, made out of the same fabric as that knot.

A sheet is only two dimensional. Our universe is not. We need a third dimension. So turn your top hat upside down and tap it with your magic wand. *Abracadabra!* A flash of light and a puff of smoke, and that rubber sheet is now a solid block of transparent rubber extending in all directions. You're standing inside of it, so let's make you a ghost so you can glide around unimpeded. Our knot is now three-dimensional, like a möbius doughnut, maybe a little silvery like a bubble underwater. It's not really made out of anything, it hasn't got a colour, and it hasn't got a surface. It's just a soliton, a travelling twisting stress that's a photon going nowhere fast, so now we call it an electron. Our electron has replaced our cannonball, and now we need a photon to stand in for that rolling marble. Let's conjure one up, and send it propagating through our rubberworld so that it passes by our electron, and let's see what happens. The photon is a transverse wave, also know as a shear wave. In mechanics a shear wave travels at a velocity determined by the stiffness and density of the medium. There's an equation for it that goes like this:

$$v = \sqrt{(G/\rho)}$$

The G here isn't a gravitational constant, but is the *shear modulus of elasticity*, to do with rigidity. It's different to the *bulk modulus of elasticity*, because it's a lot easier to bend something rather than stretch it out. The ρ is the density. The equation says a shear wave travels faster if the material gets stiffer, and slower if the density increases. You can't directly apply the concept of density to space, but in electrodynamics the velocity equation is remarkably similar:

$$c = \sqrt{(1/\varepsilon_0\mu_0)}$$

Here the ε_0 is permittivity and μ_0 is permeability. High permittivity means a material will take a larger charge for the same voltage. Barium titanate has lots of it, a thousand times as much as air, so we don't make capacitors out of air. High permeability means a material exhibits more magnetic flux when you change the charge. Iron has lots of it, wood doesn't, so we don't make magnets out of wood. Now remember that *charge is twist*, which tells you space is easier to twist if permittivity is increased, hence it's "less rigid" and $1/\varepsilon_0$ can be likened to G. And remember that *magnetism is turn*, which tells you that space will turn things better if permeability is increased, hence it's "more dense" and μ_0 can be likened to ρ. There's some marvellous similarities between mechanics and electrodynamics, though some things are flipped. For example with the piezoelectric effect you compress a material with inward pressure, you exert a mechanical stress, and the result is a voltage:

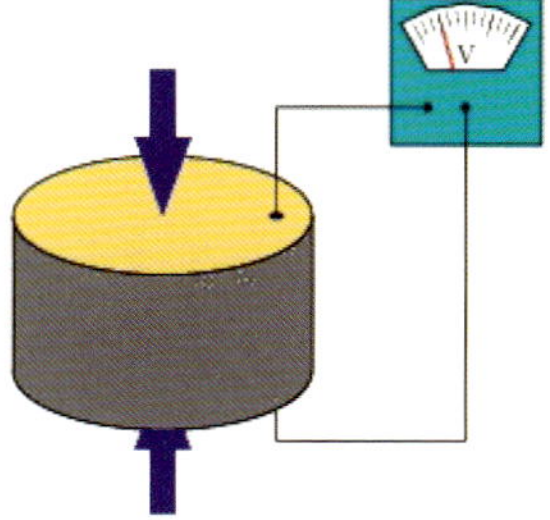

Figure 74 – Piezoelectric effect

But high voltage is called high tension, and tension is negative stress, and an electric current goes from negative to positive. Hence things are back to front. So for space, velocity doesn't increase as we approach our central stress. Instead it reduces. It reduces because the pressure is outward rather than inward. It's like pushing your fingers into a taut rubber sheet and then spreading them. The energy changes space to create a negative tension gradient, but it's easier to just call it a tension gradient. It really is like this. This is from Einstein's Leyden address of 1920:

"Mach's idea finds its full development in the ether of the general theory of relativity. According to this theory the metrical qualities of the continuum of space-time differ in the environment of different points of space-time, and are partly conditioned by the matter existing outside of the territory under consideration. This space-time variability of the reciprocal relations of the standards of space and time, or, perhaps, the recognition of the fact that 'empty space' in its physical relation is neither homogeneous nor isotropic, compelling us to describe its state by ten functions (the gravitation potentials $g_{\mu\nu}$), has, I think, finally disposed of the view that space is physically empty".

That's the truth of it. Space isn't empty, because space is full of space. Matter/energy is a stress in space that conditions the surrounding space, like a knot in a rubber sheet affects the surrounding rubber. And as to how it works for real, it's to do with vacuum impedance, which is $Z_0 = \sqrt{(\mu_0/\varepsilon_0)}$, and is measured in ohms. The symbol for ohms is Ω, you might have seen it on a circuit board. Impedance is like resistance, but for alternating current rather than direct current:

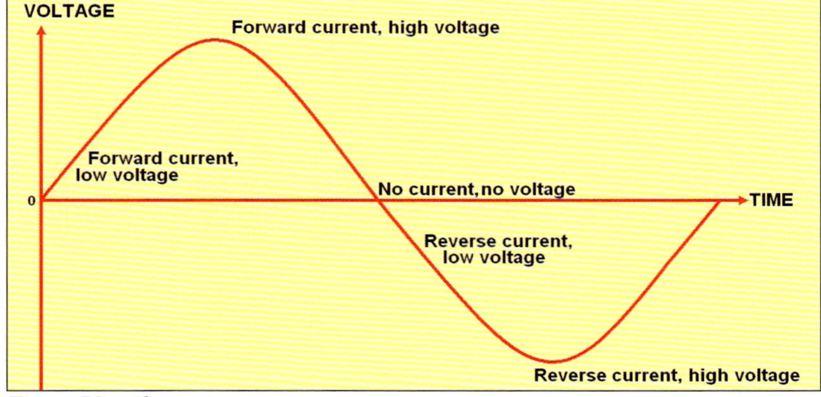

Figure 75 – Alternating current

What's interesting is that $1/\varepsilon_0 = \mu_0 c^2$. Does that look at all familiar? It ought to. It's like $E = mc^2$. And what it's saying is that $1/\varepsilon_0$ and μ_0 are like energy and mass, in that they're just two different ways of looking at the same thing. It's all about how twistable and turnable space is. The long and the short of it is that impedance is all about the *strength of space*. And space is like elastic. If it's stronger, and you're a photon, it's harder to stretch. Impedance increases as we approach the central stress. It's like piezo-resistance, the more impedance, the tougher the going. A photon is a quantum of alternating current, a higher vacuum impedance equates to more resistance, and that results in a lower velocity.

Back in rubberworld, our photon-marble is passing our electron-cannonball. We notice it veers *towards* it a little. That's because where the rubberworld stress is slightly greater, the real-world impedance is slightly higher, so the velocity of light is slightly lower. What we're seeing is a type of *refraction*. It's not quite the same as refraction through a glass block, but it's similar. It's so very similar that when we see it through our telescopes, we call it gravitational *lensing*.

Figure 76 – Gravitational lensing

Here's the crucial point: our real world is like ghostly transparent elastic containing ripples of stress, some of which are tied into knots. And we are a part of it, we are ripples and knots too. We are painted into the bulk of the rubber that is space. We stretch with it. We are made out of this insubstantial fabric. We are so totally immersed in it and so much a part of it that we cannot directly measure any change in tension, nor vacuum impedance. And nor can we directly measure any change in the velocity of light, because we can only measure the velocity of light in terms of the velocity of light. We are *made* of this stuff, along with our rulers and clocks. We can't measure the change in c locally. It would be like trying to measure the length of your shadow using the shadow of your ruler. It always measures the same, be it morning, noon, or dusk.

But we *can* infer the change in c. We can measure it from afar, by comparison. It's there in the gravitational time dilation, programmed into our GPS[31]. It's there in the Pound/Rebka experiment[32], where a photon appears to be blue-shifted at the bottom of the tower. It's there in the Shapiro Effect[33] where light takes "longer" to skim the Sun.

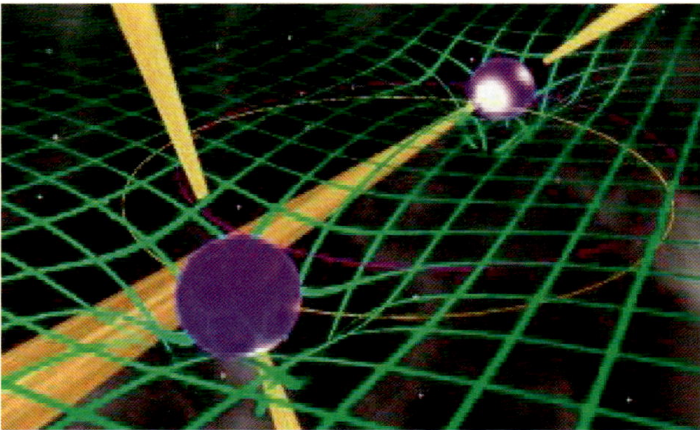

Figure 77 – Shapiro effect

We can see it as plain as day but we don't know what we're seeing. We're raised from birth with a conviction that c is set in stone, and the idea that this belief is wrong is just too much. We prefer to wonder about the mystery of gravity, instead of seeing it for the simple thing that it is. Because of course, to challenge the constancy of the speed of light would be to challenge relativity, right? Wrong. Einstein knew all this, but it's been airbrushed out of history by translation and misinterpretation. This is what he said:

"In the second place our result shows that, according to the general theory of relativity, the law of the constancy of the velocity of light in vacuo, which constitutes one of the two fundamental assumptions in the special theory of relativity and to which we have already frequently referred, cannot claim any unlimited validity. A curvature of rays of light can only take place when the velocity of propagation of light varies with position".

It's clear from the context that whilst Einstein used the word velocity, he meant it in terms of speed; a scalar not a vector. If you insist upon the strict definition of velocity wherein a change of vector direction is considered to be a change in velocity, Einstein's comment is reduced to a tautology. He would have been saying: *light curves because it changes direction.* That's nonsense. He wasn't saying that. He was telling us the speed changes, and that gravity involves a variable speed of light[34]. This means current thinking is not in line with Einstein's gravitational field[35], and all you have to do to realise this, is open your eyes and look.

Figure 78 – Plastic lens used in astronomical microlensing simulation

When we look with fresh eyes we see a powerful equivalence between special relativity and general relativity. The equivalence is this: *time dilation is direct evidence of a reduced speed of light.* It's easy to see why.

Imagine that I stay here on Earth while you travel to Alpha Centauri and back. We use a gedanken spaceship that travels at .99c and doesn't have to worry about acceleration and deceleration, and we use $1/\sqrt{(1-v^2/c^2)}$ to work out that you experience a sevenfold time dilation. We normally think of time dilation as being countered by length contraction, but this only occurs in the direction of travel. Hold up a metre ruler *transverse* to the direction of travel and it's the same old metre. Your metre is the same as my metre, and your time is dilated by a factor of seven, which means it takes a beam of your light seven times "longer" to traverse your transverse metre. Don't get confused by this. Don't persuade yourself that your light beam is following a diagonal path and has to cover a greater distance. If you do, you're looking at your situation from *my* reference frame. You must observe *your* situation from *your* reference frame. Stay in your frame. Stay in your box. Reduced to unassailable simplicity, speed equals distance over time, your metre distance has not changed, your time however has, and therefore your speed of light has also changed. *Your c is a seventh of mine.* So when you return from your year-long round trip, you find that I aged seven years, but you only aged one. You aged less because your c was less than mine, but you never noticed it at the time, because *you are made of light.*

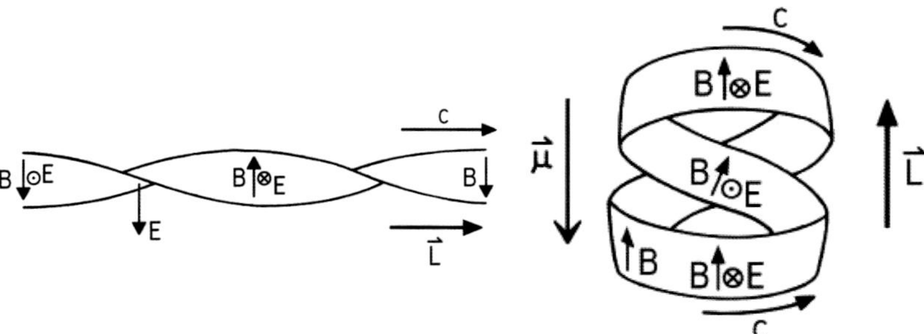

Figure 79 – Photon strip and figure-of-eight electron strip

When we move from special relativity to general relativity, we can appreciate that I could have subjected you to a high gravity environment instead of sending you to Alpha Centauri. We know from GPS that "clocks run slower" in a gravitational field, as they do when we're travelling fast relative to each other. And it's for the same simple reason: clocks clock up motion, not time, a change in c alters the motion of light, light defines your time, and that alters your seconds. The c is reduced. But you won't measure it as reduced, just as you'll never measure three ounces to the pound on the moon. We're made of light, and clocks run slower only when light runs slower. That's why the delayed Shapiro light really was delayed, because light really did travel more slowly when it skimmed the Sun. That's why the blue-shifted Pound/Rebka photon is in fact unchanged, but gravitational time dilation alters our observation of it. And that's why the GPS clocks are set to run at 10.22999999543 MHz instead of 10.23 MHz. The experimental evidence is there already, hiding in plain view, as obvious as the nose in front of your face if only you know how to look[36].

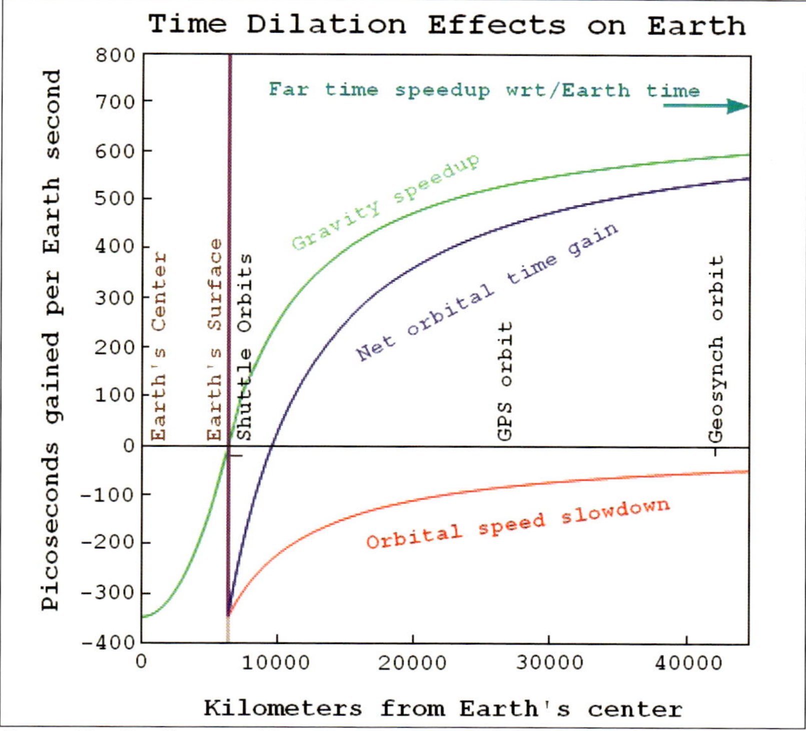

Figure 80 – Satellite time dilation

I know it's difficult to stop thinking of c as a constant. Yes it's always measured to be the same "in all frames". But when you step back to see the big picture that is the whole shooting gallery, when you look at all the frames side by side, you see what distinguishes them is the way c changes, secretly, immeasurably to somebody in the frame. It's a constant, *but it isn't constant*. The bald truth is that a gradient in c is what a gravitational field is, and when you can appreciate this, you can allow yourself the epiphany of understanding gravitational potential energy. We know that $E=mc^2$, so a cannonball sitting quietly in space represents maybe 10^{11} Joules of energy. If a planet now comes onto the scene, the cannonball will fall towards it, and just before impact will also have kinetic energy of say 10^8 Joules. Now, we ask, where did this kinetic energy actually come from? Has it been sucked out of the planet via some magical mysterious action at a distance? Let's ask an expert, somebody who was right on the money:

"That gravity should be innate, inherent, and essential to matter, so that one body may act upon another at a distance through a vacuum, without the mediation of anything else, by and through which their action and force may be conveyed from one to another, is to me so great an absurdity that I believe no man who has in philosophical matters a competent faculty of thinking can ever fall into it". [37]

Figure 81 – Newton was right on the money

That's a no from Sir Isaac Newton, and he should know. So, has it been magically extracted from some zero-point bottomless bucket? No. Has it come from the gravitational field? No, there's no free lunch from Mister Gravity, because the force of gravity is a *pseudo force*, Einstein said so. And we know already that a falling body *experiences no force*. No force in play means no energy is being delivered. So the only place the energy could have come from is the cannonball itself. And it hasn't come from its mass because mass is "invariant". So $E=mc^2$ and we've got a pile of cannonball kinetic energy that hasn't come out of the m. There's only one place left it can have come from. The c. The c up there is greater than the c down here, and there's a gradient in between. There's *always* a gradient in c when there's gravity. Yes, the gradient might be very small. But it isn't negligible. If you think it is, as per the Principle of Equivalence, you've thrown the baby out with the bathwater. An accelerating frame with no tidal force isn't the same as standing on the surface of the Earth. There's always a tidal force in a real gravitational field. The gradient has to be there. There can be no Uniform Gravitational Field, it's a contradiction in terms, a hill with no slope, a *flat* hill. Because without that local gradient that's there in your box, *things don't fall down.*

Let's go back to rubberworld and look at *why things fall down*. But it's time we did a *reverse image* and made the rubber the ghost. Now you're back to normal again, take a look at that electron once more. It's a travelling stress localised because it's going round in a circle. Stick this ring of light that looks like this: ○ into a real gravity gradient, caused by a zillion other electrons and protons some distance off. What's going to happen? Let's divide the circle into four flat quadrants and make it very simple:

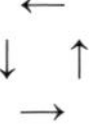

Flash, take a picture. At a given instant we have a photon travelling down like this ↓. There's a gradient in c from top to bottom, but all it does is give the photon a slight blue-shift. A little while later take another picture. *Flash.* Now the photon is moving like this → and the lower portion of the photon wavefront is subject to a slightly lower c than the upper portion. So it bends, refracts, curves down a little. Later it's going this way ↑ and gets fractionally red-shifted, and later still it's going this way ← and bends down again. These bends translate into a different position for our electron. The bent photon path becomes electron motion. The electron falls down:

The reducing speed of light effectively bleeds motion out of the component photon and into the electron. But only half the cycle got bent, so only half the reduced c goes into kinetic energy aka relativistic mass. Where does the other half go? The other half goes into rest mass, the "invariant" mass. The invariant mass *varies*. It increases. The Pound-Rebka experiment tells us this. There's a gravitational blue-shift because the c is reduced so it looks like there's more energy in the photon at the bottom of the tower. And when we consider the electron to be a photon configuration where the energy appears as mass, it would look like there's more mass in the electron. But there's no energy being added, that's just a scale change falling out of the clear blue sky, so here's your free lunch:

Figure 82 – There is no free lunch

Now you can understand why light is deflected twice as much as matter. Gravity is not some magical mysterious action-at-a-distance force. It's not curved spacetime. There are no hidden dimensions, there's no blizzard of gravitons sleeting between the masses. There's no energy being delivered, so gravity isn't a force. It's just the tension gradient that balances matter/energy stress. That's why there can be no gravity waves per se. Stress and tension walk hand in hand, two sides of the same coin, action and reaction, they grip each other. There can be no waves of tension without stress. If two circling neutron stars lose energy, there has to be stress accompanying the outgoing tension, just like a photon. So LIGO, the Laser Interferometer Gravitational-Wave Observatory will never see gravity waves using light. We need to use clocks.

Figure 83 – LIGO

You can also understand why gravity is so weak. The matter/energy stress is concentrated at one location, but the tension gradient that balances it extends across the universe. It's the Dirac Large Numbers Hypothesis[38] made simple. And once you see it for what it is, gravity is simple too:

Gravity is an extended tension gradient opposing matter/energy stress, wherein the speed of light varies resulting in gravitational time dilation and attraction through refraction.

A NOTE ON NEWTON

Newton nearly had it. He feigned no hypothesis, but he did pose queries, proposing an explanation of gravity that most people don't know about. He didn't know about impedance and talked of density, but it's close enough. He knew it was all down to light. That's why he focussed on Opticks:

"Are not gross bodies and light convertible into one another? ...Doth not this aethereal medium in passing out of water, glass, crystal, and other compact and dense bodies in empty spaces, grow denser and denser by degrees, and by that means refract the rays of light not in a point, but by bending them gradually in curve lines? ...Is not this medium much rarer within the dense bodies of the Sun, stars, planets and comets, than in the empty celestial space between them? And in passing from them to great distances, doth it not grow denser and denser perpetually, and thereby cause the gravity of those great bodies towards one another, and of their parts towards the bodies; every body endeavouring to go from the denser parts of the medium towards the rarer?" [39]

Newton knew full well that gravity was a refraction, and why an apple fell down. He knew that gross bodies and light could be converted into each other. And if you can change matter into light, changing base metal into gold had to be a piece of cake. Hence the interest in alchemy. He was ahead of his time, so far ahead that the world wasn't ready.

We live in a "hard light" world. Shades of Ace Rimmer. We're made of light, the speed of light isn't constant, and that's what gravity is. There is no fundamental time, so spacetime isn't fundamental, and nor is curved spacetime. Because that's what it *does*, not what it *is*. Because spacetime is space with motion through it, and everything we thought fundamental is slipping through our fingers. The last man standing is the photon. And so it goes: *the world is painted in light.*

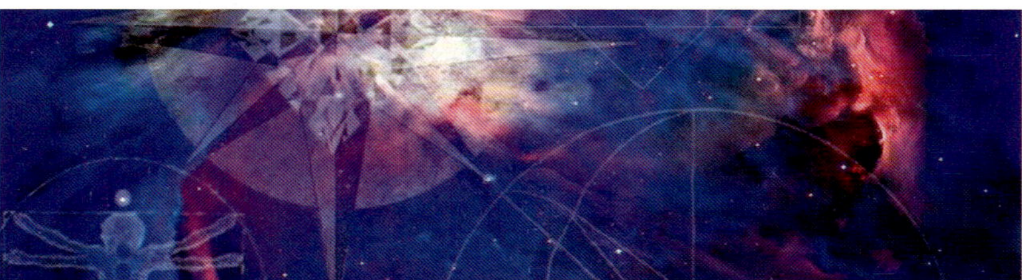

Figure 84 – The world is painted in light

SPACE EXPLAINED

OK, now we're getting somewhere. Energy is a stress volume moving through space at c. Mass is a measure of this energy when it's going nowhere fast. Charge is the twist that keeps it in place. Time is a relative and cumulative measure of motion, light defining our distance and time. And gravity is the tension that opposes stress, a gradient in the impedance of space rather than curved spacetime, a variable c that we can never measure locally.

But what the heck is space? We often think of it as *nothing*, but actually it's impossible to imagine nothing. Nothing is a tricky concept that people take for granted. Close your eyes and try to imagine nothing. What colour is it? Blue? No. It can't *be* nothing if it's got a colour. How heavy is it? A kilogram? No, it can't *be* nothing if it's got a mass. How big is it? A metre across? No, it can't *be* nothing if it's got a size. Here's a picture of it: Did you get that? Here it is again: Got the picture? There isn't any picture. When you dig down to what nothing is, you find it has no properties at all, and you can't hold it in your mind. It slips between your fingers. You can't imagine nothing because it's just not there. Nothing doesn't exist.

Space isn't nothing. Space has got a size, so space definitely isn't nothing. But that's all it's got. It's a one-trick pony, and the only trick is distance. Yet just like Google, it's pretty darn brilliant.

Figure 85 – Brilliant one-trick pony

Let's bring it closer to home and have a closer look. Have you ever travelled on the London Underground? On certain stations the platform curves a little, and over the Tannoy you will hear those immortal words: *mind the gap*. This gap has no colour and no mass, but it does have a size. It's got a property, and the "possession" of this property makes all the difference. Possession is nine tenths of the law, and a gap has distance. It's the very last property that makes it the thing that it is. It's a gap, it has distance, it *is* distance, it *is* the space between the platform and the train. When you think of what a gap is, you will appreciate that this "no-thing" thing *only* exists because the two objects exist. It's like defining a hole. A hole is something that only exists by virtue of the thing it's a hole in. My negative carpet is a hole in my living room carpet. In itself it's not actually there, but we treat this lack of something as if it's something that has independent existence. We say it exists, we call it a something, but at the same time it's a nothing. It's something and nothing. And if you had a hole in nothing, that would be something.

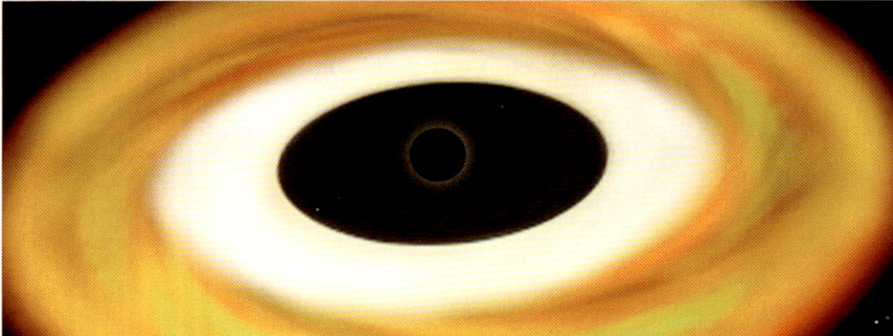

Figure 86 – Non-spinning black hole

So, space is distance. Fair enough. How do we measure distance? Yes you can stretch your hands out or use a ruler, maybe a metre rule. But that metre rule is in turn defined using something else. In the end we measure distance using light. Because light defines our distance:

The metre is the length of the path travelled by light in vacuum during a time interval of 1/299792458 of a second...

We also use time to measure distance, particularly when we're moving. But we know that time is just a relative and cumulative measure of motion. Time is defined by light too:

Under the International System of Units, the second is currently defined as the duration of 9,192,631,770 periods of the radiation corresponding to the transition between the two hyperfine levels of the ground state of the caesium-133 atom...

We know that radiation is light, as are the internal atomic motions that caused it, because matter is made out of light. Our electrons are made of light, our atomic clocks are made of light, and our measuring rods are made of light. We are made of light, and everything around us. It's a totally immersive situation. If we woke up tomorrow morning to find that c had halved, we wouldn't notice it at all. The speed of light would still be measured at 299,792,458 m/s, and the observable universe would still be so many light years across. But everything would be half the size it used to be. This is why I said *the world is painted in light*. It would be like bringing the screen closer to the projector. The people living and breathing in the projection wouldn't see the difference.

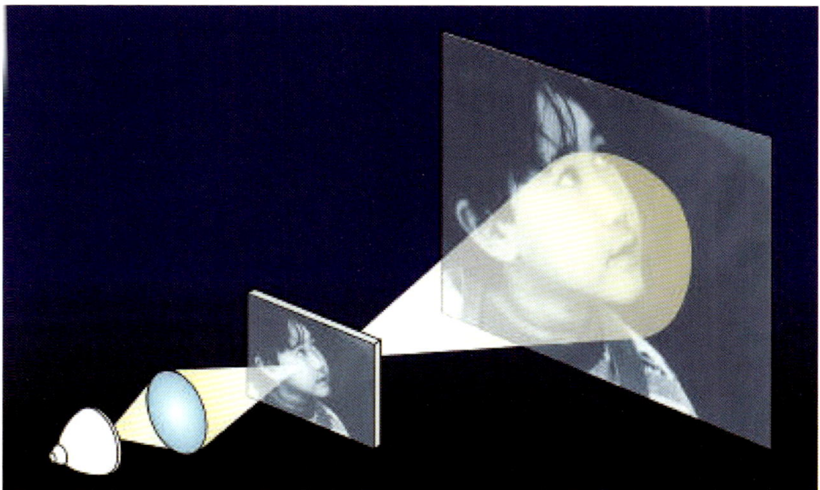

Figure 87 – Projection

And the moot point is this: the world is painted in light, and *so is the canvas*. We have to remember this when we think about space. The distance between two objects is defined in terms of light, and the objects themselves are made out of light. So if you take away the light, it's very difficult to think in terms of distance any more. How big is the space between two objects if the objects aren't there, and the measuring rod isn't there either? It doesn't parse. It does not compute. It's a contradiction in terms. It's like asking the colour of up. To get beyond this we have to take a look at light. A real close look. We know it's a travelling stress volume, energy moving from one place to another, and we normally talk about it as a propagating electromagnetic effect.

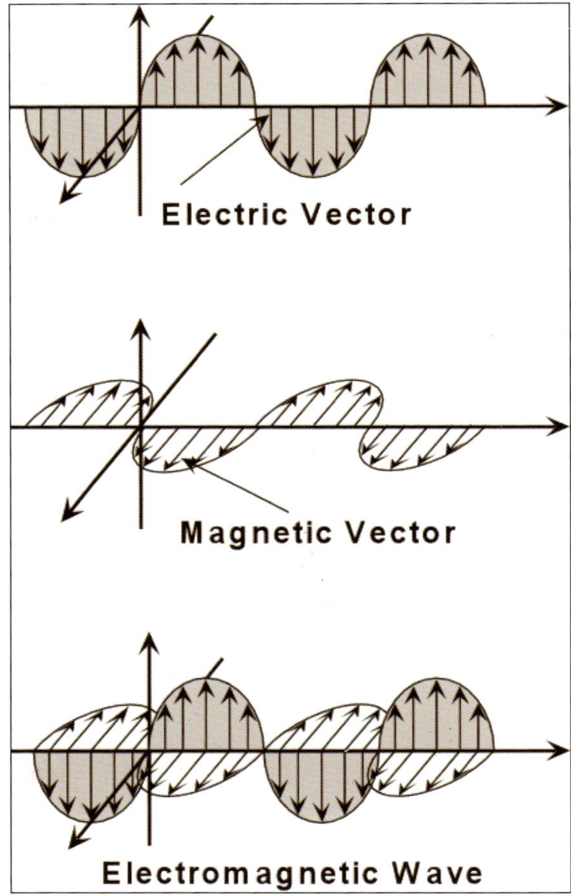

Figure 88 – Light

We have a mental picture of an electric field varying with a sine wave frequency, and an orthogonal magnetic field similarly varying. But a photon has no charge, and is not magnetic. Electromagnetism is what we get when we employ pair production to convert a photon into an electron and a positron, thereby creating mass and charge. It's charge that gives us an electric field that we view as a magnetic field when we travel through it. You can make charge via pair production, and you can destroy it via annihilation. That means charge is not fundamental, just as mass is not fundamental. That means the electromagnetic field that comes hand in hand with charge isn't fundamental either. If the world was a sea of photons there wouldn't *be* any electrostatic attraction. There wouldn't *be* any magnets. So we can't really say that the photon is the mediator of the electromagnetic force. That's getting things back to front, like everything else to do with electricity. The photon is more fundamental than the force it mediates. Because it's energy, and whilst you can destroy charge, you can't destroy energy. No, it isn't the electromagnetic field that's fundamental. What's fundamental is *polarization*.

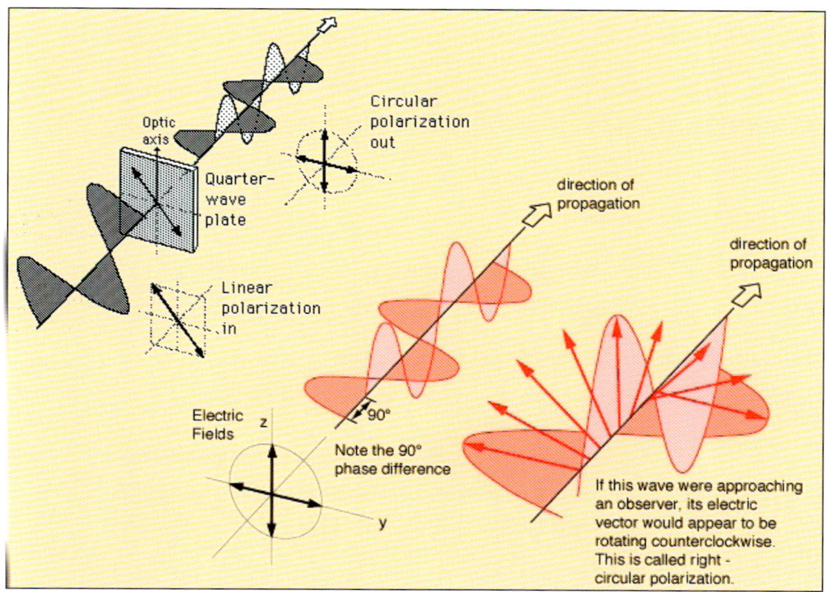

Figure 89 – Circular polarization

The fundamental thing that the photon has "got" is *polarization*. By this I don't mean the *direction* of the polarization. I mean the polarization itself. I'm talking about the vacuum polarization of space here. Note that a photon is often thought of as being a simple up-and-down thing, something vertical and flat. It isn't like that. If you shake a skipping rope up and down, you can see an up-and-down transverse wave, but think of your skipping rope embedded in a rubber sheet. Things are happening on either side. When the rope goes up like this ↑, the rubber on either side is moving like this ↗↖. The arrows are vectors, and you can decompose them to say that the rubber is moving like this →↑←. You can take it further by imagining your skipping rope is firmly fixed into the surface of a rubber block. Now when you shake it up and down, you can visualise how the bulk of the rubber has a whole arc of vectors in its motion. To go all the way you have to imagine a wave in your rubber block when there's another rubber block on top of it, and no skipping rope. Here we're dealing with a pressure wave, a push rather than a pull, so the vectors are like this ←↑→.

It works in three dimensions, and that's why an electric vector ↑ can be decomposed into component vectors ↖↗ as evidenced by a quarter-wave plate. It's easiest to think of a circularly polarized photon as a clock at ten to two, only the minute hand has been bent towards you. It's similar on the downside where vector ↓ is decomposed into ↙↘ and the right-hand vector is bent away from you. As a result the net electric vector rotates as the photon goes by. People often think the photon is rotating, but that's misleading, it's just orthogonal vectors out of phase. What's even more misleading is that the electric vector isn't the direction of motion. The electric field is a *twist* field, the electric vector is telling you the twist. And that's why polarizing filters don't work like the skipping-rope and picket-fence analogy. The photon that's polarized up and down gets through the *horizontal bars*, not the vertical bars.

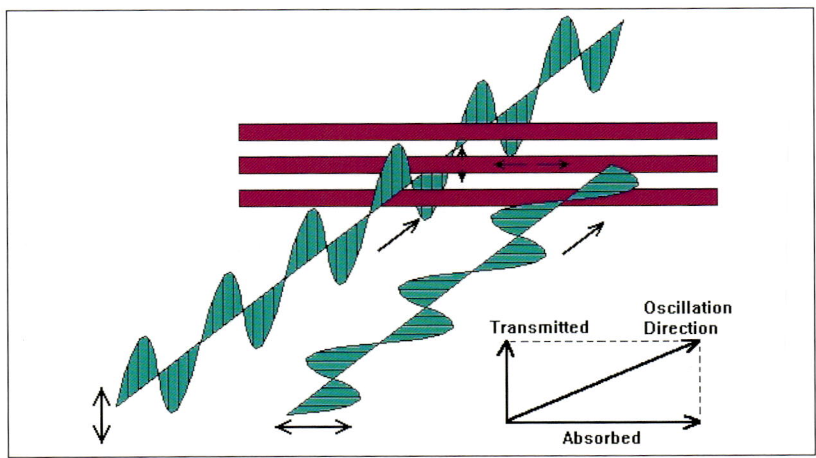

Figure 90 – Polarizing filters

All of this is why there's more to vacuum polarization than the direction of an electromagnetic oscillation. Strictly speaking, the photon isn't even oscillating. It's like an ocean wave, if you paced it in a helicopter you wouldn't see any oscillation at all. When we think of ocean waves we tend to think of breaking waves, but it's better to think of oceanic swell. These are big gentle sine-wave undulations that can be as big as a ship. Bigger, because they aren't just a bulge on the top of the sea. A test particle on the surface moves like this ᵔ as a swell wave passes. If the test particle is down in the water it moves like this ᵕ, if it's deeper down it moves like this ᵕ, and so on. A swell wave is a pressure wave, and as it passes through the bulk of the sea the ocean bulges. Vacuum polarization is akin to this. It isn't the direction of the bulge, you have to think of the bulge itself.

Figure 91 – Swell waves

I can relate polarization to the poles of a battery. This time the battery isn't f at. So what does it have? Yes, it has charge. That's a measure of the total number of electrons available to flow from the negative pole to the positive pole, because electricity is back to front. But it has something else too. Across its poles. We call it polarization. We also call it *voltage.* You probably think of voltage as the pressure that drives an electric current around a circuit. But it isn't like that. It can't be, because pressure is stress and *stress* x *volume* = *energy*, and *voltage* x *charge* = *energy* too. Charge isn't volume, so voltage can't be pressure. So what is it?

Get an elastic band. Hold it vertically, stick a pen through the top end and grab the bottom end firmly. Start twisting using the pen. Keep on twisting until the elastic band starts kinking. OK, you know I said charge is twist? Well, you have basically "charged up" your elastic band. It's only an analogy, but it's a good one. Now, give it some voltage. *Do something to the elastic band to increase the energy stored within it.* There's not a lot of options. This is what you do:

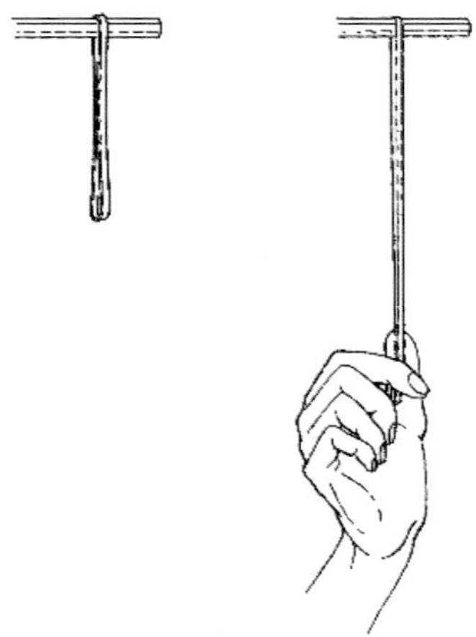

Figure 92 – Elastic band

What you have to do is pull the two ends apart. You increased the energy in the elastic band. You added some "voltage". Now look at it carefully. What's different? I'm not asking you to feel the tension. Just look at it and *see* what's different. It's obvious, so obvious that you might even miss it. What's different is the distance between the two ends. *You increased the distance.* If you think about a 511 keV photon twisted to make an electron, the charge is what it took to twist the photon into a möbius doughnut. It doesn't take any *extra* energy to make an electron. It's still only 511 keV. But now we've got charge, an electromagnetic field. A charged particle exerts an electromagnetic force on another charged particle. A photon does not. We say $F=ma$ because the electromagnetic field is a *force* field. It's the electromagnetic *force*, it's there because the charge is there. It's like a tetherball, there's a force in play but it isn't doing any work, so there's no extra energy involved. That's why charge isn't energy. Because charge is twist, and when you boil it right down, charge is *force*.

You need a force and a distance to have energy: *force* x *distance* = *energy*. So if *charge* x *voltage* = *energy*, and charge is force, that means voltage is *distance*. That's the stunning simplicity of it. Voltage is polarization. Light is a travelling polarization, travelling through space. It travels in one direction, and the polarization is in an orthogonal direction. The polarization isn't the direction of polarization. It's the change in distance in the orthogonal direction. And if you want to twist some space, you have to change a nearby distance. That's why charge is twist. You increase a distance in space, and it makes space twist.

A whole new geometrical concept is now coming together: the electromagnetic field is in no way fundamental. It's merely twisted space that appears to be turning space if we're moving through it. To twist this space, we must change a nearby distance. Doing this is what we call current. We do it for a distance, which is what we call voltage. And when we change a distance in nearby space, we cannot readily create a gap in space. Something very strong holds space together. And furthermore, when we change a distance in nearby space, we have to make up for the change in distance in all the *rest* of space. The stress that is local has to be balanced by a tension extending across the universe.

A photon of light is energy travelling through space. The stress rides with the tension, like a wave in a rubber mat. The two go hand in hand, they're two sides of the same coin, like action and reaction. That coin is energy. And as it passes through space, there is more space, because *the change in distance is a distance*, and it makes for an increased volume. Hence space and energy are also two sides of the same coin. That very first definition *energy is a volume of stressed space* was telling us something important:

Space is a one-trick pony, and the only trick is distance. The photon is energy, and is a rippling change of distance that makes for more volume. A change of distance is a distance, so space and energy are the same thing.

This one-trick pony really is just like Google. And it *is* pretty darn brilliant. As to how brilliant, we shall see.

EINSTEIN'S HEROES

Here's a few excerpts from Einstein's last interview[40], with Isaac Bernard Cohen, the science historian who translated Newton's Principia:

"She conducted me to a cheerful room on the second floor at the back of the house. This was Einstein's study. It was lined on two walls with books from floor to ceiling and contained a large low table laden with pads of paper, pencils, trinkets, books and a collection of well-worn pipes. There was a phonograph and records. Dominating the room was a large window with a pleasant green view. On the remaining wall were portraits of the two founders of the electromagnetic theory – Michael Faraday and James Clerk Maxwell... Einstein said that he had always admired Newton. As he explained this, I remembered those striking words in his autobiographical statement following a critique of Newtonian concepts – "Newton, forgive me"... "It is important to know", he went on, "what Newton thought and why he did certain things". We agreed that the challenge of such a problem should be the major motivation of a good scientific historian. For instance, how and why had Newton developed his concept of the ether?"

Einstein had a picture of Faraday on his wall, next to the picture of Maxwell. He used to have a picture of Newton too, but it fell off its hook:

Figure 93 – Isaac Newton

Newton, Faraday, and Maxwell were pioneers of modern physics, and they were Einstein's heroes. They weren't his only heroes of course. Another of Einstein's heroes was Lorentz, who was something of a father figure as you can read in Walter Isaacson's book *Einstein: his life and universe*[41]. There's more, I wouldn't presume to know them all. All I can do is tell you that you should read the books and read the history, because the history of science is littered with brilliant men, and you should read the detail, and see what they said for yourself. You'll gain an appreciation of the slow pace of progress, and how the baton is passed. There's more, because you'll understand that heroes aren't super-heroes, that they're human, and that makes them all the more heroic. They've got hopes and fears and other things too. They even have heroes, because we've all got heroes, like I've got heroes, like Feynman and Dirac and Einstein too. There's more because Einstein's heroes[42] are my heroes, including Newton, Faraday, and Maxwell. And when you read what they say, they come alive, like they're talking to you across the centuries. That's when you sit up and pay attention. Because that's when you realise that all this stuff has been kicking around for a long long time:

"And first, I suppose, that there is diffused through all places an aethereal substance, capable of contraction and dilatation, strongly elastick, and in a word much like air in all respects, but far more subtle". [43]

That was Sir Isaac Newton, in a letter to Robert Boyle in 1679. He was talking about elastic space. He moved away from this idea for a while, but later moved back to it. And it wasn't just Newton who thought along these lines. Here's another excerpt from Einstein's Leyden address of 1920:

"When in the first half of the nineteenth century the far-reaching similarity was revealed which subsists between the properties of light and those of elastic waves in ponderable bodies, the ether hypothesis found fresh support. It appeared beyond question that light must be interpreted as a vibratory process in an elastic, inert medium filling up universal space. It also seemed to be a necessary consequence of the fact that light is capable of polarisation that this medium, the ether, must be of the nature of a solid body, because transverse waves are not possible in a fluid, but only in a solid. Thus the physicists were bound to arrive at the theory of the "quasi-rigid" luminiferous ether, the parts of which can carry out no movements relatively to one another except the small movements of deformation which correspond to light-waves".

There's more because once you've got a handle on how things hang together, things start jumping out at you. Here's another excerpt::

"Of course it would be a great advance if we could succeed in com-prehending the gravitational field and the electromagnetic field together as one unified conformation. Then for the first time the epoch of theoretical

physics founded by Faraday and Maxwell would reach a satisfactory conclusion. The contrast between ether and matter would fade away, and, through the general theory of relativity, the whole of physics would become a complete system of thought, like geometry, kinematics, and the theory of gravitation. An exceedingly ingenious attempt in this direction has been made by the mathematician H. Weyl; but I do not believe that his theory will hold its ground in relation to reality".

That's Einstein talking about Faraday and Maxwell when he's talking about *the small movements of deformation which correspond to light waves.* And Hermann Weyl:

Figure 94 – Michael Faraday, James Clerk Maxwell, and Hermann Weyl

Faraday was a real hands-on guy with a total grasp of his subjects. He was an experimentalist, he was brilliant, and he invented so much. You really should read up on his lectures[44]. They're vivid, full of demonstrations, an unforgettable show and tell rather than a dry old talk. Here's what Einstein said about him:

"He conceived these fields as states of mechanical stress in an elastically distended body (ether/space). For at that time this was the only way one could conceive of states that were apparently continuously distributed in space. The peculiar type of mechanical interpretation of these fields remained in the background - a sort of placation of the scientific conscience in view of the mechanical (Newtonian) tradition of Faraday's time".[45]

Maxwell is thought of as more theoretical and mathematical, doing the equations that formalised Faraday's practical work. However he also did crucial experimental work on the kinetic theory of gases, and on light. For example he invented colour photography and presented it at a Royal

Institution lecture in 1861. Do read up on his Dynamical Theory of the Electromagnetic Field[46] and note his electric *elasticity* equation E = 1/ε D. People generally don't know about this, and don't generally know that Oliver Heaviside recast Maxwell's quaternion equations in vector form, and totally destroyed the insight of elastic displacement. They never seem to hear about this:

"We can scarcely avoid the conclusion that light consists of the transverse undulations of the same medium which is the cause of electric and magnetic phenomena."

Weyl was the mathematician who many consider to be the inventor of gauge theory. He likened gauge transformation to something akin to a change in the scale of railway tracks. The Weyl transformation[47] is a local rescaling of the metric tensor, and the latter is used to measure distances and angles in space. It's all about changing distances in space, but when your ruler is part of that space, you can't measure it directly. Instead, like a flatlander confronted by a rumple in his two-dimensional world, what you experience is a force. Weyl was really on to something with this, but sadly he wasn't one of Einstein's heroes. Instead Einstein knocked him back, and Weyl was left dead in the water for ten years until his material was picked up by quantum theory. *That* was Einstein's greatest blunder[48].

And that brings me on to a puzzle that's been rumbling on for years. It started way back in the seventeenth century. Huygens wrote his Treatise of Light, talking about waves, whilst Newton talked about corpuscles, or particles. The wave/corpuscular debate went in Newton's favour for a hundred years, then swung to Huygens for a hundred more. But it wasn't quite so cut and dried, because Newton's corpuscles had "fits of easy transmission and easy reflexion" caused by the vibrations of his elastick aether. That's in Opticks query 21. His corpuscles were not simply billiard balls. His was the first hint of wave/particle duality.

PARTICLES EXPLAINED

The word "particle" is responsible for a whole pile of problems in physics. It comes with this mental baggage that makes people think of shiny little billiard balls, all in different colours. Yes, a picture is worth a thousand words, but when push comes to shove, sometimes it's the wrong picture:

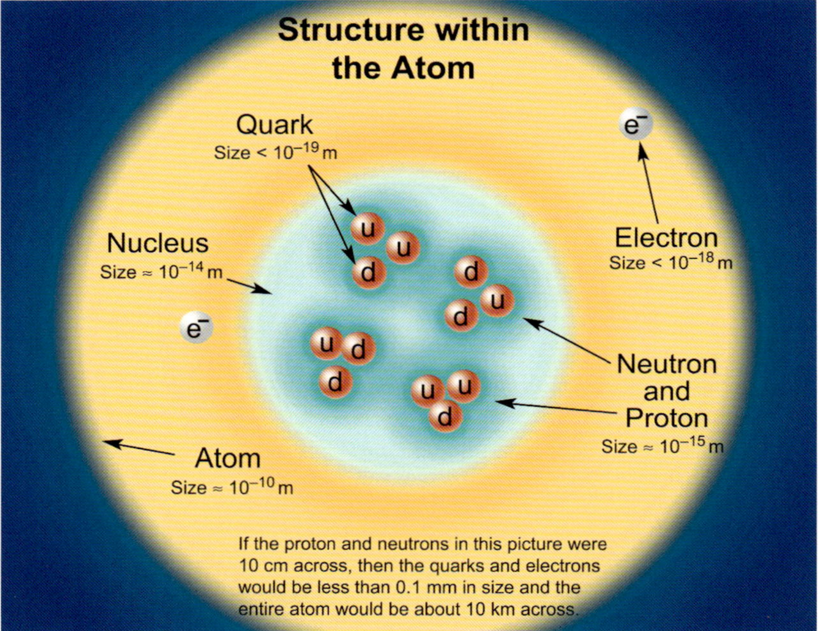

Figure 95 – Structure within the atom

Particle physicists will swear that they don't really think in terms of billiard balls. But some of them do. They just don't know it. Worse, they talk about point particles of zero size, creating impossible mathematical infinities. They do this to make the mathematics easier, forgetting that something that has no size can't have angular momentum, oblivious to Dirac, oblivious to the fact that they've just made a mystery out of spin. Then some of them express amazement that their nitty-gritty specks can be in two places at once, and start dreaming up weird ideas like parallel universes to explain it. It's easy to be in two places at once. Put your hand on your heart: *you are here*. Now put your other hand on your head: *you are here too*. Now look at your hands. See? You are in two places at once. And *everything* that exists is like that. Because if something really was of zero size, it just wouldn't be there. You might beg to differ and show me the point of a knife, but lose the knife that's behind it, and where's the point? Nowhere. No, you have to get these billiard balls and point particles right out of your head, because it just isn't like that in the subatomic world. It's just geometry in action, and once you understand it, it's simpler than you ever thought possible.

By now you should have gained a rough idea of the electron. It's a soliton, a möbius doughnut, a knot in space. It's a trivial knot, the simplest you can make, and it can be drawn like this:

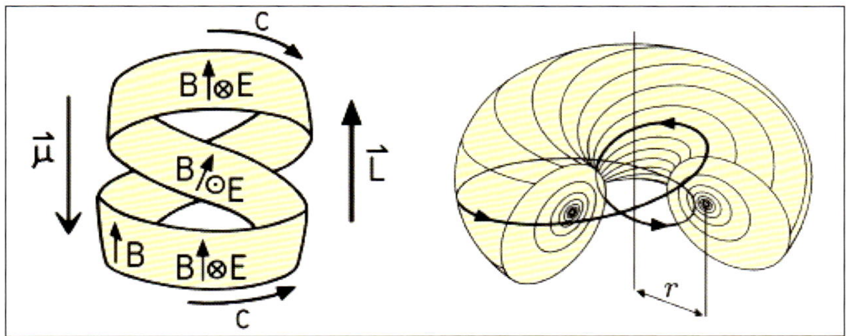

Figure 96 – Strip electron and möbius doughnut electron

But you have to remember that neither picture is perfect. It isn't a paper strip like a figure of eight, and it doesn't look like a doughnut. This knot has no surface, and no substance too. A slice through the doughnut might be drawn like a slice through an onion, but the onion rings don't stop, and there are no rings. We aren't dealing with a solid object, and knowing this, we start to think in terms of fluid. The electron is a photon going round in circles, and a photon is like an ocean wave. It doesn't have a surface and it doesn't have a substance, because the ocean's got the surface and water's got the substance. The photon is a travelling stress, and when it's tied up as an electron it's like a whirlpool in the middle of the ocean. There is no outer edge, so there is no surface, so there is no size. You might say the whirlpool is ten feet across, but that's just the hole in the middle, and for our electron, there's no surface there either. Some say the electron is less than 10^{-18} metres across, but that's just the figure they got down to in scattering experiments+ without finding internal structure. It's like looking for a cannonball in the middle of the whirlpool, and it's just not there. And truth be known, nor is the whirlpool. The fluid analogy is useful, and yes, it gives you a feel for the Pauli Exclusion principle because two waves can ride over one another whilst two whirlpools can't overlap. But it doesn't get to the heart of it, because a fluid is made out of particles, like a solid is made out of particles, and a gas is made out of particles. And what's at the heart of it is this: space isn't made out of particles, particles are made out of space. Intangible space is what everything is made of. Tangible things, like electrons and atoms and clocks and buses, everything under the Sun, and the Sun as well. They're made out of energy, and energy is space, but we can't nail it down like we can nail down a ripple in a rubber mat. If we want to hold it in place, there's only one tool in the box.

The first clues as to how are readily available in everyday life, where we see many objects that we think of as being intangible. For example, think of a knot in a string. You can't take the knot out of the string and hold it in your hand, because the knot in itself it isn't a tangible thing. The knot itself has neither surface nor substance, because these properties belong to the string. The crease in my pants is in a similar class. It's a one-dimensional discontinuity, a line a little under a yard long. It has no colour or mass, but there's no mystery to it. It's just the crease in my pants, sharp, like the blade of that knife. A fold is something similar, as is a twist. When these things are motionless, we talk of geometry and topology, but when they're in motion, then we talk of *action*. Consider a shout. A shout is a thing, but can you measure its width? How about its length? Can you pin down its location? The answer to all these questions is no, because a shout is an *action*. You understand what it is, there's no mystery to it, you don't think of a shout as being made up of shout particles, but you do think of it as an intangible thing that you can't get hold of. A kick is another action. It doesn't weigh anything, it doesn't have a colour, or a surface or a shape. It has no substance, a kick isn't something you can hold in your hand. Whilst it might therefore seem to be intangible, a kick from a pony will challenge that view.

All this geometry and action makes a difference, and it's a big big difference. They make things *more* tangible. A wave in the surf is an action that can knock you and your boyfriend flat on your back, laughing and screaming with salt water up your nose. You can't grab hold of it, but it can grab hold of you. Be it a kick or a wave, an action is a halfway house between tangible and intangible, and there is a way to go all the way. There's a way to solve the mystery of how intangible space can be employed to construct tangible matter, the particles from which we're made, from which everything is made. It all comes back to photons.

A photon is an action, like a kick is an action. Unlike a shout, which is a longitudinal wave in air, a photon is a transverse wave in space. It's a ripple, a shimmy, and now there's no mystery to it. Because space is a one-trick pony, and the only trick is distance. A photon is a travelling stress moving through the volume of space, but what it really is, is a rippling *change of distance*. It's usually shown as a wavelike variation of the electric field accompanied by an orthogonal variation in the magnetic field, both being aspects of the electromagnetic field:

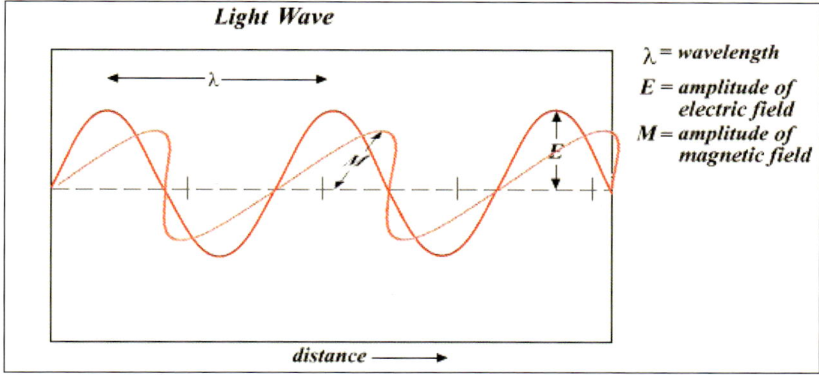

Figure 97 – Light wave

But I've talked about the electromagnetic field as twisted space, wherein the electric field is a "twist field", and the magnetic field is a "turn field" view of the self-same thing whilst in motion through it. In order to effect a twist in space, we must change a distance in nearby space. For an electromagnetic wave to exist, there must be some transient change of distance in the local space. This will twist the surrounding space, effecting a simultaneous turning motion. Hence the photon is a propagating pulse of space in space moving through space. When we tie down the change of distance into the twist in space that we call charge, we're creating the photon configuration that we call an electron.

It's like tying a knot in a string, only there is no string, just a distance variation travelling a knotted path so making for all-round twisted space. The photon is an action, and an action cannot be an action without motion, just as a kick cannot be a kick if the leg is still. The action is still there, travelling continuously at c, but in a tight twisting loop. It's *going nowhere fast*, with zero net motion with respect to the observer, so presenting momentum as inertia. This photon configuration is made out of massless intangible light, but it's got mass, and it's tangible too. It's stable because it's tied in a knot, and the möbius doughnut is such a knot. It's the simplest knot, a trivial knot. A photon tied in such a knot now exhibits charge, mass, and spin ½ along with zitterbewegung[49] jitter. So we call it an electron. And that's all it takes to make matter out of energy.

To get a real grasp of this, you need that hands-on experience. You need to play with paper. Start with a strip of paper that will represent a photon. Mark it with an E on the top left and top right. Draw a flattened U-shaped sine wave across the length of it, to represent the electric field over one photon wavelength. Now mark it with an M on the bottom left and bottom right, and draw another flattened sine wave to represent the magnetic field, but draw it the other way round like a flattened arch. Note that the fields are orthogonal and reach their maxima together, so you should imagine your paper is creased along its length into an L-shaped profile. It's like you're looking at the photon wavelength marked with the lambda <—λ—> in figure 97 on the previous page, but from a forty-five-degree angle below. You end up with something like this:

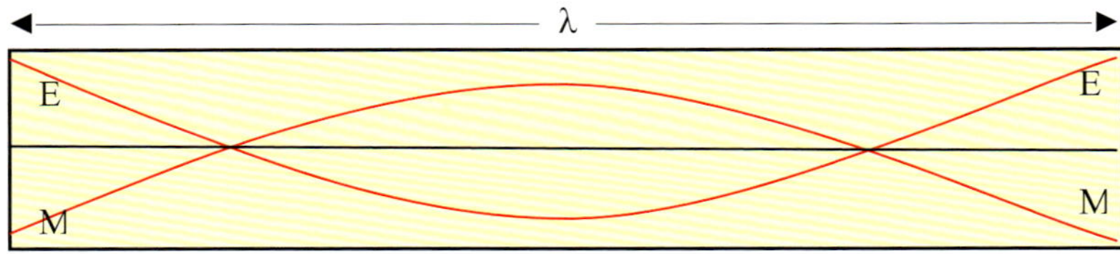

Figure 98 – Paper strip

Turn the paper over sideways and repeat the drawing, ensuring that the Es and the Ms on one side match the reverse. Now loop the paper around and twist it to make a möbius strip, then hold the ends together. Note how the E lines up with the M, and the M lines up with the E. That's the crux of it. In the electron, the electric field *is* the magnetic field and vice versa. That's why it's stable. It grips itself. The twist is the turn and the turn is the twist. It twists *and* it turns, the twist *is* the turn, you can't distinguish them. The travelling change in distance that causes the twist is following a turning path *caused by itself*. Note however that the electron is a spin ½ particle, the component photon is wrapped round twice. You might think you ought to slide the two ends of the möbius over one another and keep on going until you've got a double-wrapped möbius. But it's not like that, because *the lines on the paper are the twist*. Hence you can let go of the möbius and fold your strip in half, first lengthways, then vertically. Then make a simple loop, and that's the möbius doughnut, or at least a flattened version of it. You started with a photon, but now you've got yourself an electron. Or a positron. Because if the twist goes one way it's an electron, and if it goes the other way it's a positron. But whichever way it goes, you have to go round 720° to get back to where you were, which is why it's like a möbius strip. Möbius strips are like that. Cut a möbius strip in half all the way round the middle, and you don't get two möbius strips. You get a circle with a barrel-roll twist plus another one for luck, and the circumference is twice the length it was. It's similar with photons. Start with a 1022 keV photon and use pair production followed by annihilation to produce two 511 keV photons, and each is twice the wavelength of the original. Clever stuff this three-dimensional geometry. Cut things in half and they're twice as big.

As to how it works in practice, imagine you're very small and you're standing in a three-dimensional cubic lattice representing space, just as a photon is passing through. I've just pressed the *freeze-frame* button, so you can see that the vertical line-elements of the lattice are stretched up above your head, rather like an ocean swell. The photon has pressure and volume like the ocean wave but there's no surface involved, hence the line-elements are also stretched beneath your feet. The degree of stretch diminishes in a lemon-like sinusoidal fashion in front of you and behind, and also diminishes on either side, where the horizontal line-elements are bent to accommodate all the vertical stretching. We really need computer modelling to see it properly with wave trains and packets and everything else, but it looks something like this:

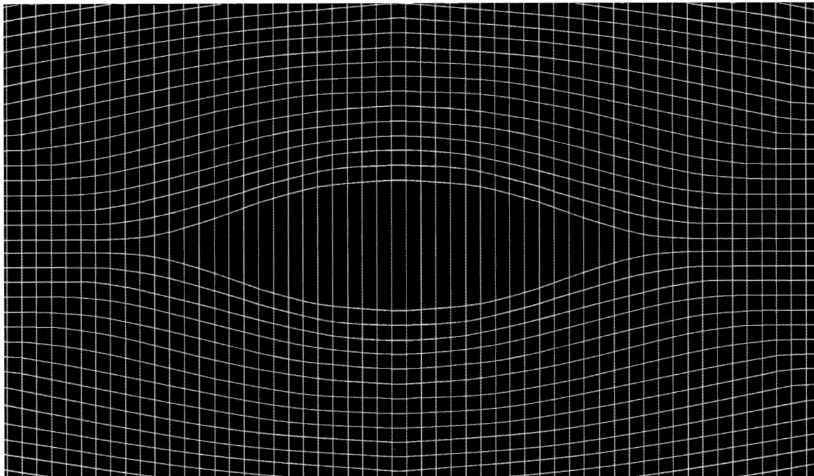

Figure 99 – Photon lattice

The twisted lattice around you is the transient electric field. It looks a little different to the picture of the light wave with its sinusoidal electric and magnetic fields. But it's the same because all we've got is a distance variation, and you can't have a negative distance. If you look closely at figure 99 you can see there's no twisting to the left of the photon, none in the middle, and none to the right. Now imagine you're on the right of the photon, and I've just pressed the *play* button. As the photon passes you'd feel the twist increase to a maximum, then reduce as the middle of the photon passed you by, then go negative and then to zero as the photon went totally by. That's why the electric field varies in a sinusoidal fashion:

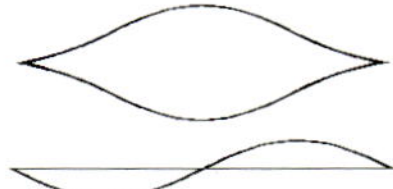

And of course as all this twisting occurred you'd see the line elements actually bending, effecting a turning action this way and that. That's the magnetic field, and like the electric field it's just another way of looking at it, and it's been and gone in a flash because that photon is zipping along at the speed of light.

The thing to remember about all this is that the photon isn't really some lemon-shaped pulse that you can actually draw, because the photon is just where the bent lattice lines are, and there is no lattice. There's just no surface to it, and there isn't any size. Because it isn't an object, it's an action, and any shape we draw is just a depiction of what the action does as it passes you by. It's a distortion, an extended entity that is nothing more than distortion, but draw it we will. If the photon has a longer wavelength, we draw the sinusoidal lemon shape longer, with less intensity, and shallower angles to its shoulders:

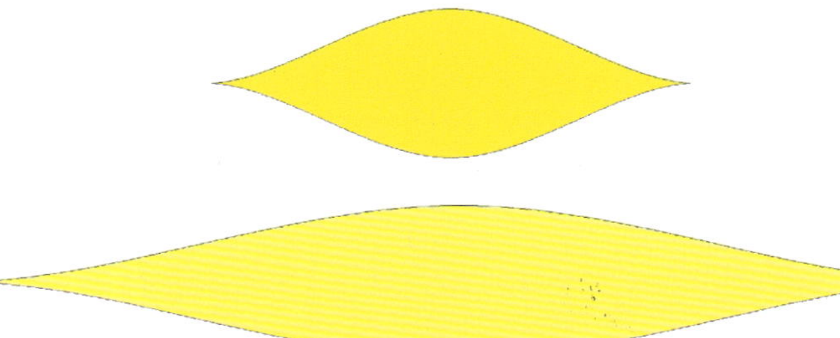

Figure 100 – 1022 keV photon and 511 keV photon

When you've got an energetic gamma photon of 1022 keV you can create an electron and a positron. All you need is a nucleus. It acts like a breakwater, or like an axe. When the photon encounters it, the photon is split into two. It isn't cut vertically down the middle like you'd cut a lemon in half. You can't cut a spacewarp in two like this: . It's just distortion. It isn't a solid object. And if you cut it horizontally across the middle like this: , that doesn't leave a symmetrical shape. The only way to split a photon is to split it lengthways through the centre like you'd split a log. Then suddenly the two sides of your log are twice as long as they used to be, and they're going different ways. Get it just right and the two halves change direction so much that they bend right round and over. They twist and turn, and eventually the bend becomes so pronounced it loops right round on itself. Again we need computer modelling to see it properly, but if a photon was flat like paper it would look something like this:

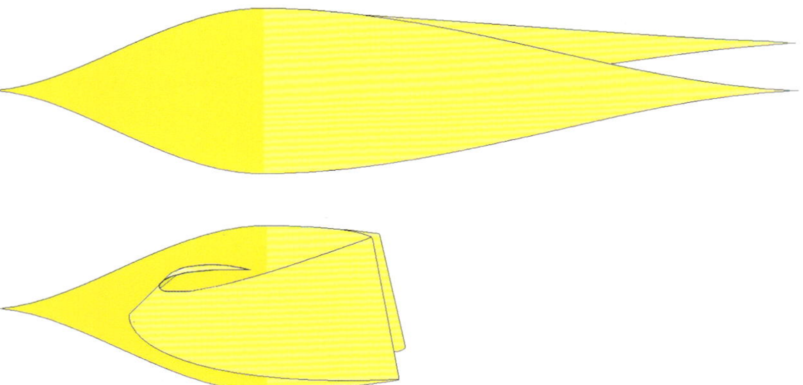

Figure 101 – Photon splitting, and photon undergoing pair production

The original 1022 keV photon peels into two 511 keV photons and each forms a stable eddy. In truth there has to be a little more than 1022 keV so there's enough energy to send the electron and the positron flying apart, otherwise they'd annihilate immediately. But assuming we can catch them up and look at them, we'd see that the electron is going this way: ℧, and the positron is going this way: ℧. We'd see that each 511 keV photon changed direction so abruptly that it encountered its own twisting space, caused by itself. This causes a further change in direction, and is repeated continuously. At 511 keV the two aspects of the twisting turning variation in distance match exactly and lock together. The distance variation causes space to twist, and this causes the distance variation to travel in a turning path. It turns round twice to get back to its original position and orientation, so it's like a möbius strip. The result is a stable soliton at 511 keV, and at that value alone. This 511 keV can be viewed as a measure of the "strength" of space, representing the energy that goes into creating a trivial knot in that space. We call the trivial knot an electron or a positron, depending on the chirality. It's made of light, but we call it matter. It's a wave, it's a particle, it's a wave and a particle, and it's simple when you see it:

The photon is a wave of distance variation. We can tie it in knots to make particles with mass and charge. The electron is a trivial knot, the positron is its mirror image. So matter is just space with a twist that's tied in a knot.

Once you understand wave/particle duality and what particles really are, you can finally blow away all the spooky mystic magic that's been masquerading as quantum mechanics for eighty years. The word quantum means "how much". That's all. And all the mystery evaporates like Scotch mist in the searing summer Sun once you understand the quantum of quantum mechanics.

QUANTUM MECHANICS

Quantum mechanics evolved during the early years of the twentieth century in response to the black body problem, where the energy distribution of light from a cavity oven was charted as a hump rather than increasing continuously with frequency. Max Planck got the ball rolling in 1901, establishing that there was some kind of cutoff that prevented what's now called the ultraviolet catastrophe. The emission of light seemed to be subject to some kind of discreteness or packeting. The energy was quantized in its relationship to frequency, such that E=hf always applied, where h is Planck's constant of action.

But nobody knew why, and progress was slow. Einstein wrote his photo-electric paper in 1905. It was called *On a Heuristic Viewpoint Concerning the Production and Transformation of Light* [50], and said the ability of light to knock electrons out of a metal plate depended on the energy of individual quanta rather than the intensity of the incident beam. This meant that light itself was quantized, not its emission and absorption, and thus light was particle-like in its nature. But many were not convinced. Planck himself wasn't persuaded until 1911, and then the first world war put everything back. It wasn't until 1918 that Planck won his Nobel Prize, followed by Einstein in 1921. Then Arthur Compton proved Einstein right in 1922 via Compton scattering, and light really did seem to consist of particles rather than waves. Hence in 1923 when Louis De Broglie proposed that matter is made of waves, it was totally revolutionary. De Broglie's hypothesis made a massive difference to the earlier Bohr model of the atom. Instead of billiard-ball electrons orbiting like planets, you move to an integer number of electron waves round the nucleus, like dérailleur gears:

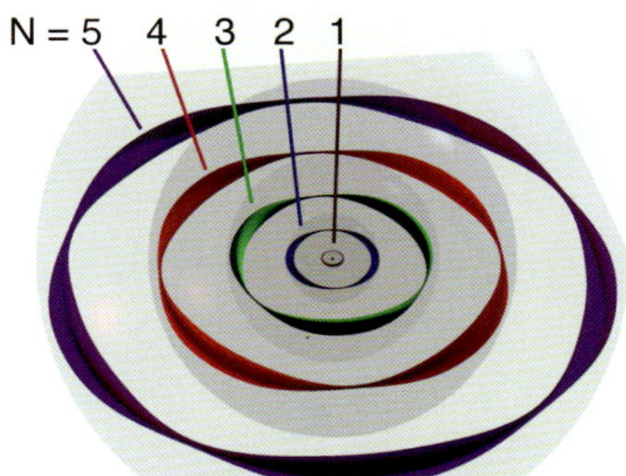

Figure 102 – De Broglie atom

Next came Heisenberg and Born with matrix mechanics in 1925, a purely mathematical approach which gave no picture of the underlying phenomena. Then in 1926 Schrodinger came up with his wave equation, and the name "photon" was coined by Gilbert Lewis. A year later De Broglie's hypothesis was confirmed experimentally by Davisson and Germer via electron diffraction. Shortly thereafter Dirac was coming on-stream with his relativistically invariant form of Schrodinger's wave equation, predicting electron spin and the existence of the positron. It was a time of great excitement, but De Broglie got sidetracked by pilot waves, Schrodinger got talked out of classical wave mechanics by Lorentz of all people, and everything went pear-shaped.

The Solvay Congress of 1927 was perhaps the greatest meeting of great minds ever. There's Albert Einstein centre stage, in the middle of the front row, sitting next to Hendrik Lorentz on our left. That's Marie Curie next to Lorentz, then Max Planck. Sitting behind Einstein there's Paul Dirac and Arthur Compton, looking like brothers. Just to the right is Louis de Broglie, and on the back row behind them, there's Erwin Schrodinger.

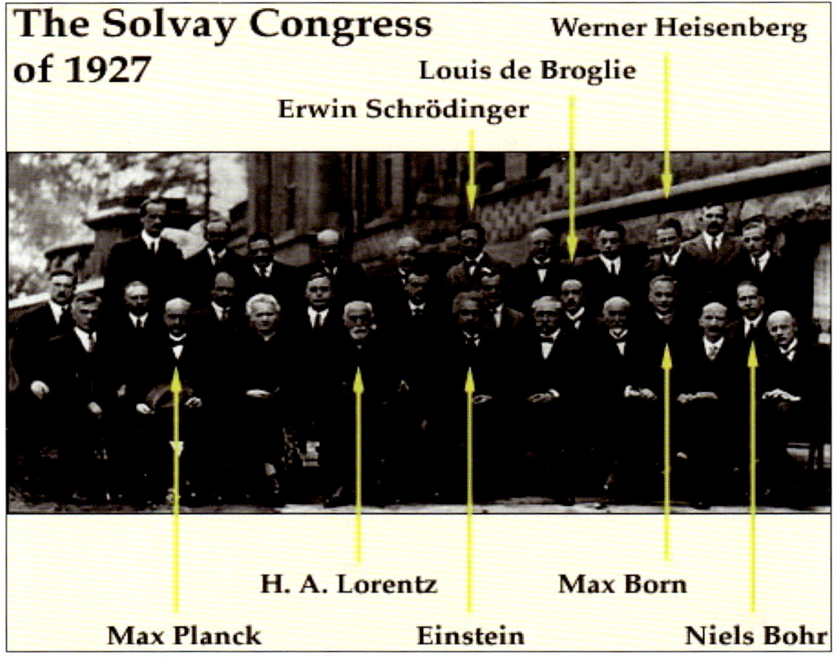

Figure 103 – Solvay congress 1927

The Solvay Congress of 1927 was a great meeting of great minds, but it wasn't a meeting of minds. Because whilst Einstein and Schrodinger and Planck were in one camp, Bohr and Pauli and Heisenberg and Born were in a different camp. The wrong camp won, and the probabilistic Copenhagen Interpretation set in. It comes with the point-particle idea of the collapse of the wave function, leading to the unscientific speculation that nothing exists until it is measured, along with the unprovable conjecture of parallel worlds that we can never ever see. It also comes with the conviction that nobody can ever understand the subatomic world, that spin is mysterious, and that all attempts to grasp a picture of reality are doomed because they're classical and outdated.

And yet it would appear there was a simple answer all along: *the wave function is the particle*. It describes where the extended entity is, not where a point particle can be detected. The "many paths" and "virtual particles" of QED reflect this extended nature of a wave/particle photon or electron. The photon is a distortion. When it travels in a straight line there's an element of that distortion on either side of the line, reducing as you move away from the median. That's why the photon seems to take paths other than the direct path. The "time reversed electron" is indeed a positron, but due to opposite chiral motion rather than "travelling backwards through time".

The subatomic world *can* be understood. Understanding electromagnetism in terms of geometry means electrical permittivity can be viewed as the twistability of space, and magnetic permeability can be similarly viewed as the turnability of space. They're akin to mass and energy in being two different measures of the same thing. We combine them in the simple expression $c = \sqrt{1/\varepsilon_0\mu_0}$ to give the speed of light. We also combine them as impedance $Z_0 = \sqrt{\mu_0/\varepsilon_0}$, impedance being resistance to the alternating-current photon. These two combinations are intimately related, and conceal a pleasant surprise.

We know that photons propagate through space at c. Now consider the local space whilst a photon is propagating through it. How quickly does the photon energy cause an extension in this space? If the photon is a low-frequency radio wave, the extension will proceed gently. It's sinusoidal, it will start slow, then increase in rate, then slow again before the extension reaches a maximum, and then it will follow the reverse pattern. If however the photon is a high-frequency 1022 keV gamma photon, the extension is more rapid. But *it cannot be infinitely rapid*. If we assume the maximum rate of extension is c, and consider that the extension is occurring for a lesser duration in a high-frequency photon, we can see that regardless of frequency, the extension is always the same. The photon *amplitude* is always the same, it's a physical distance, and there it is in every picture of the electromagnetic spectrum you've ever seen:

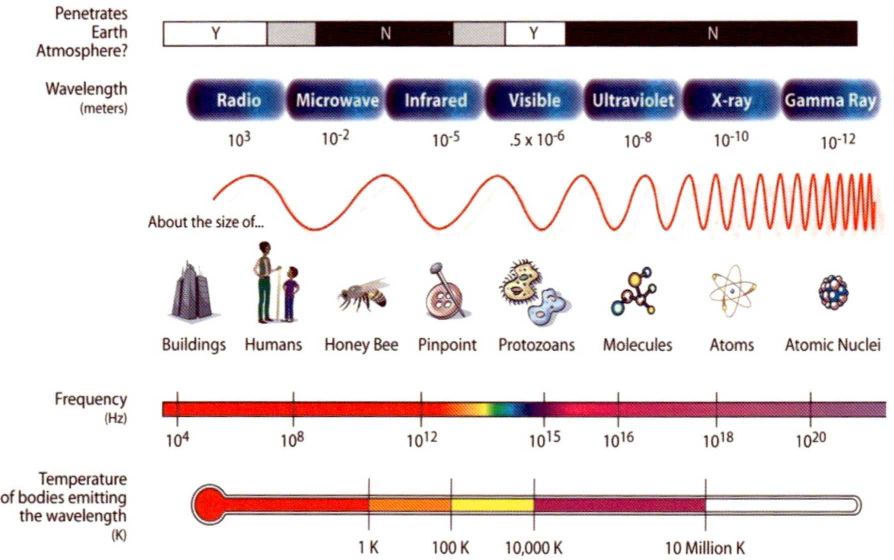

Figure 104 – The electromagnetic spectrum

To tie a photon in a spin ½ trivial knot where the wavelength represents two turns round a circle, only one wavelength will suffice. It's 2π times the diameter, and that diameter is the common photon amplitude. *This* is why the electron and the positron come in one size only, equivalent to 511 keV. The wavelength of a 511 keV photon is 2.426×10^{-12} metres. If we divided this by 2π we would say that the overall crest-trough amplitude of photons is 3.86×10^{-13} metres.

The energy of a photon can be written as E=hf where h is Planck's constant of action, and f is frequency. Different photons have different frequencies, different energy/momentum, different action, different "kick". But action is momentum multiplied by *distance*. Planck's constant of action isn't just a constant of action. The common photon amplitude transcends Planck's constant. It's a *distance*. It's the cutoff value that explains the black body problem. It's the quantum of quantum mechanics:

The common photon amplitude is a spatial extension of 3.86 x 10^{-13} metres, and is the quantum of quantum mechanics. The wave function doesn't describe where a point particle can be found, it describes where the extension is.

Once you understand all this, there's more you can understand. You can understand all the particles. Not just the photon and the electron, but neutrinos and protons, and neutrons and quarks, and everything else.

Figure 105 – The particle zoo

MORE PARTICLES

Let's take a look at some more particles, starting with the muon. The muon is considered to be a lepton like the electron. It's in the same "family", along with the tau. There's also the associated electron neutrino, usually just called the neutrino, along with the muon neutrino, and the tau neutrino. This family also includes a full complement of antiparticles such as the positron, the electron antineutrino, and the antimuon.

The electron weighs in at 511 keV, the muon at 105,658 keV. The muon is therefore 206.77 times more massive than the electron. It has the same negative charge, the same half-integer spin, plus it has an antimuon partner, just like the electron has an antielectron partner in the positron. So at first glance the muon seems very similar to the electron. But it isn't. It isn't stable, and that makes all the difference. It usually decays in about two nanoseconds into an electron, an electron neutrino, and a muon neutrino. It's clearly not merely some super-size trivial-knot electron. It can't be. Space has a strength. There's only one mass/energy for the trivial knot con-figuration. The muon isn't it, and it isn't stable, therefore it's not a knot. It must represent some different configuration, and to understand it, we must understand the neutrino, and to understand the neutrino we must understand the proton and the neutron. Consider the illustration below:

Figure 106 – Trefoil knot

This is the trefoil knot. It's the next simplest knot after the trivial knot. Follow the loops carefully and note that two trips around the trefoil are required to return to the starting location and orientation. This feature matches the trivial-knot electron, and suggests a spin ½ entity with unit charge. Now follow the loops again starting from the bottom left and continue twice around the knot, noting the directions at the crossing points. Since each crossing point is encountered twice, omit alternate crossing points to only consider crossings *over* rather than under. The crossing-point directions are: up, omit, up, omit, down, omit. Does that ring any bells? Up up and down? Why, this is a *proton*, and it consists of two up quarks and one down quark. The loops are the quarks. They can't be separated and maintained as quarks. That's why we've never seen a free quark. And the tensile strength of the trefoil can be treated as gluons mediating the strong force. Imagine the trefoil to be made of strong elastic, and pull at any one of the three loops. The stretching force required will increase as you increase your pull. This is why we've never seen a free gluon. Gluons are "messenger particles", but they're accounting units rather than genuine particles. We've never actually seen them because they don't actually exist.

We know the proton mass/energy is 938,272 keV, which says the component Compton photon wavelength is 1.3214 x 10^{-15} metres. This value is much smaller than the common photon amplitude of 3.86 x 10^{-13} metres, so this knot is tight. Imagine trying to tie a knot an inch wide in fishing line a foot thick, and you begin to appreciate just how tight this knot is. This knot is *strong*. Whilst proton decay has been proposed, it has never been observed, and no small wonder. If decay were possible however, the mooted products are thought to be a neutral pion and a positron. A neutral pion decays rapidly, usually into a gamma-photon pair, but less often into an electron, a positron, and a gamma photon. In some geometrical respect, we can therefore view the proton as an electron combined energetically with two positrons, fusing three möbius-loop photons into one very tight configuration with a positive charge. Two twist one way and one twists the other, so the net twist is the same as the positron. Hence the unit charge.

Moving on to the neutron, we know that it's charge-neutral, but its magnetic moment demonstrates that concealed charge is present. We also know that a neutron is unstable outside the nucleus, and undergoes Beta-minus decay in circa fifteen minutes, becoming a proton, an electron, and an antineutrino. We further know that within an atomic nucleus, a proton can be converted into the more massive neutron via Beta-plus "decay" with the release of a positron and a neutrino. There's a symmetry here, a geometric equivalence, and it's important. Forgetting about energy and charge for a moment, we can say:

$$Neutron \equiv Proton \ \ + Electron + Antineutrino$$
$$Proton \ \ \equiv Neutron + Positron + Neutrino$$

It's important because we're talking geometry, and the geometrical difference between a proton and a neutron can in one way be viewed as an electron and an antineutrino, and in another way as a positron and a neutrino. In some sense:

$$Electron + Antineutrino \equiv Positron + Neutrino$$
$$Electron \ \ \ \ \ \ \ \ \ \ \ \ \ \ \ \ \equiv Positron + Neutrino - Antineutrino$$

This suggests that an electron is somehow equivalent to a positron plus a neutrino less an antineutrino. Now, I've previously said that an electron is the same as a positron "going the other way", like a mirror-image möbius strip. So what geometrical transformation do we need to perform on a positron to convert it into an electron?

Consider a paper möbius strip that represents a positron. We can undo the möbius strip to lay it out whilst maintaining the twist. We've removed 360 degrees of turn, and can depict our action thus: o → ∪ → —. Now we can re-form the möbius strip to create a mirror image of the original, not by reversing our action, but continuing our action in the same turning direction. We can depict it thus: — → ∩ → o. We've applied two 360 degree turns to convert our positron into an electron. We can now look again at the expression *Electron ≡ Positron + Neutrino − Antineutrino* to make a deduction:

$$Electron \ \equiv Positron + Neutrino - Antineutrino$$
$$Electron \ \equiv Positron + \ \ \ \ \ 360° \ \ \ \ - \ \ \ \ -360°$$
$$Electron \ \equiv Positron + \ \ \ \ \ 360° \ \ \ \ + \ \ \ \ \ 360°$$

This matches our simple experience with the paper strip. To convert a positron into an electron we applied two 360° turns. *We applied two neutrinos.* It's a 720° rotation, it's the Dirac String Trick, and it's telling us what a neutrino actually is. A neutrino is a lepton like an electron. An electron is a twisting turning loop. Charge is twist. A neutrino has no charge. It has no twist. It's just a turn. A neutrino is a 360° turn. It's just a loop. And yet like a photon it travels. The conclusion is clear: *a neutrino is a running loop.* We can draw an analogy with a rolling wave, or better still a bullwhip. The picture on the left represents a photon, the picture on the right represents a neutrino. Now imagine the bullwhip in action.

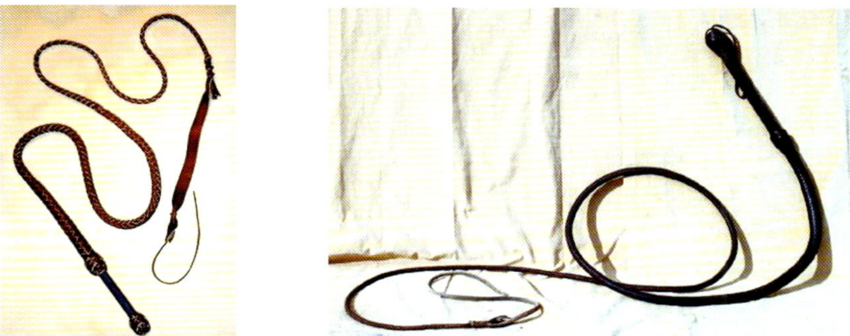

Figure 107 – Photon bullwhip and neutrino bullwhip

The helicity that we call neutrino "spin" tells us the direction of the running loop. In a neutrino the left-handed spin is a rotation around the running loop that is opposite to the direction of propagation. The antineutrino is the same running loop, but *going the other way*. The neutrino or antineutrino travels, and if it travels slower than light it has mass. That's because our new view of mass tells us that mass is merely a measure of the amount of energy that is not moving in aggregate with respect to the observer. It's a sliding scale: if the energy is moving at c like a photon, we observe none of the energy as mass. If the energy is "going nowhere fast" like an electron, we observe all the energy as mass. Ergo if a neutrino is travelling at a little less than c, it has a little mass. Should its speed vary for any reason, such as a weak interaction, its mass will also vary. We can envisage an analogy wherein a neutrino is made of slender spring steel, and as it bounces down the road it encounters friction so the loop momentarily tightens or becomes multiple loops, like a coil. The neutrino can thus oscillate, and an electron neutrino can look like a muon neutrino.

Returning now to protons and neutrons, we can add one twisted loop to the trefoil proton, representing an electron. This counters the spin ½ twist around the trefoil and sets the net charge to zero. We then add an untwisted loop to represent the antineutrino, and so make a neutron. The exact configuration is debateable, but it should change the trefoil such that alternate crossing points would read up, down, and down, yielding one up quark and two down quarks. We started with a three-way symmetrical trefoil, and we've added a twist and two turns. No way is the result ever going to be some three-way symmetrical stable configuration. In simple terms the extra loops represent a half-hearted half-hitch, and it comes undone in circa fifteen minutes. It's like an unbalanced tyre. That neutron wobbles itself apart.

Returning now to the muon, we recall that it usually decays in about two nanoseconds to create an electron, an electron neutrino, and a muon neutrino. This tells us what we need to know. The muon equates to an electron plus an electron neutrino and a muon neutrino. A neutrino is a loop, or multiple loops. Hence the muon is an electron with additional loops. It exhibits the same charge and so has the same twist. But it exhibits 206.77 times as much turn and so mass. This muon is not some möbius doughnut. It's a möbius coil. I'm not sure exactly what it looks like, but I envisage a helical tracing of a toroidal path, with the tau being the same again sweeping out a spherical shell.

Picking another unstable particle at random, consider the B–sub meson. During its picosecond life the B–sub meson oscillates between matter and antimatter because of a continuous chirality flip, like a twisted paper loop and its mirror image. You can make them on your kitchen table. They're plain loops, there's no need to twist the ends before you stick them together. You make a loop, and then you add the twist by turning one side over by 180°, and giving it a pull. You can transform one loop into the other by undoing the 180° turn and re-applying it to the other side. After a little practice it becomes a satisfying hands-on action. You can set it down in one configuration, then twist and pull again and set it down in the other configuration:

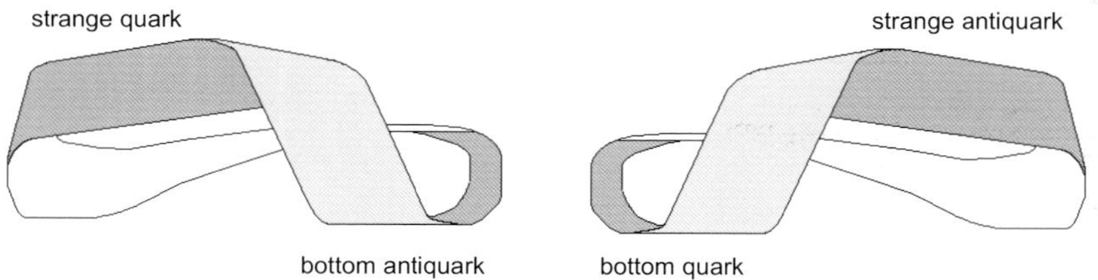

Figure 108 – B-sub meson oscillates between particle and antiparticle

There's more particles of this ilk, a lot more, be they baryons with three quarks, mesons with two quarks, or anything else. There's the neutral lambda particle, a little more massive than the neutron, but with a lifetime of only 10^{-10} seconds. So it counts for less. The charged version comes in at twice the mass but lasts for only 10^{-13} seconds. Then there's the neutral sigma particle, but the lifetime is a mere 10^{-20} seconds. There's the delta, the omega, and the cascade, and it's the same story. Check out a list of baryons and you'll see what I mean. The only stable baryon is the proton and its antiproton partner. It gets worse for mesons. The heavyweight upsilon lasts for 10^{-20} seconds, and the rho lasts for 10^{-23} seconds. There are no stable mesons. None at all. And the reason is simple. All these particles might be loop configurations, but they're not knots. They're hadronic debris, the product of collisions and other destruction events, and because they're not knots, they decay rapidly into photons and neutrinos and electrons and their antiparticles. Yes sometimes there are intermediate steps, but the intermediate decay products are just more of the same, and the final result is always the lightweight stable particles we're already familiar with. Not quarks, and not anything else. So the unstable particles aren't in the same league as the stable particles. They aren't fundamental, and neither are quarks.

And that's enough to obtain a new picture of particle physics, where the various configurations of three-dimensional knots and loops exhibit a satisfying degree of fit:

The electron is a trivial knot, a turn and a twist. The proton is a trefoil knot, three turns and a twist. The neutron is a proton plus a twist and two turns. The neutrino is a turn, a mere running loop, and muon and tau neutrinos have more loops, as do the muon and tau themselves. The antiparticles are mirror-image knots that go the other way, and the unstable particles are not knots, so they always come undone.

THE FINE STRUCTURE CONSTANT

Originally thought of as the ratio of electron orbital velocity in the Bohr atom as compared to the speed of light, the fine structure constant is the ratio of the strength of the electromagnetic force as compared to the strong force. In addition it's the ratio of the energy required to push two electrons together from infinity to some given distance, as compared to the energy of a photon with a wavelength 2π times that distance. It's a dimensionless constant called alpha, usually given as $\alpha = e^2/2\varepsilon_0 hc$, where e is the charge of the electron, ε_0 is the permittivity of space, h is Planck's constant, and c is the speed of light. It takes the value of circa 1/137.

As to why, it's to do with kissing numbers, a billiard-ball term. If electric fields were one-dimensional, an electron's electric field would extend in only two directions, so one electron would exert half its electric field on half the electric field of another, like this: $\leftarrow\!\circ\!\rightarrow$ $\leftarrow\!\circ\!\rightarrow$. Hence if you pushed them together they'd "couple" with a quarter of their total field strength, and the fine structure constant would take a value of 1/4.

If however electric fields were two-dimensional, you could push electrons together like pennies to form a hexagon. Each electron would repel another with a sixth of its electric field in any given direction, and the fine structure constant would take a value of 1/36.

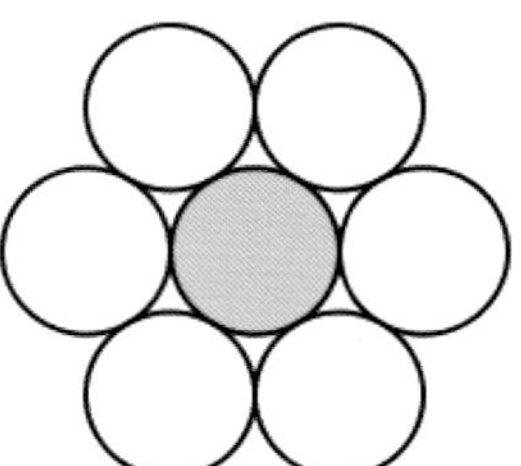

Figure 109 – Kissing in 2 dimensions

Scale it up another dimension and you can push twelve spheres around a central sphere. Each electron repels another with about a twelfth of its electric field in any given direction, and the fine structure constant takes a value that's more like 1/144. However it isn't quite 1/144 because the spheres don't quite fit. They don't fit snugly like the pennies, there's gaps between them. To appreciate this, it's best to think of an icosahedron, with a sphere centered on each of the twelve vertices. Trigonometry tells us that if each edge is 1 unit long, the radius of one large sphere just enclosing the icosahedron would be $\sqrt{(10+2\sqrt{5})}/4$ or 0.9510565163.

Figure 110 – Kissing in 3 dimensions

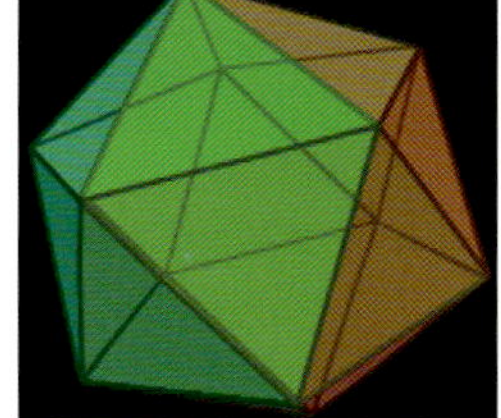

Figure 111 – Icosahedron

This means for our twelve spheres to touch, the central sphere has to be smaller. But it isn't, and that's why they don't fit snugly. There's a gap between the surrounding spheres of circa 5%. If you were the last sphere to join the icosahedron you'd be able to see more than a twelfth of the central sphere on account of this gap. In similar vein an electron "sees" more than a twelfth of another electron when it repels it with its electric field. Each electron repels another with circa $1/11.7^{th}$ of its electric field. You have to multiply this by itself because $1/11.7^{th}$ of one electric field is working against $1/11.7^{th}$ of another, so the combined coupling factor is circa $1/137^{th}$.

Note however that the fine structure constant is a *running* constant. It increases at higher energies, and is circa 1/128 at the 80.4GeV mass/energy of the W boson. We can understand this by thinking about a Fibonacci spiral. This has an exponential twist which matches the electron's electric field. The degree of twist diminishes with distance, just as the electric field diminishes in line with the inverse square law. If we imagine two Fibonacci spirals made of spring steel wire, one made of fine wire, the other made of heavier-gauge wire, we can grasp that the finer spiral is easier to twist than the heavier spiral:

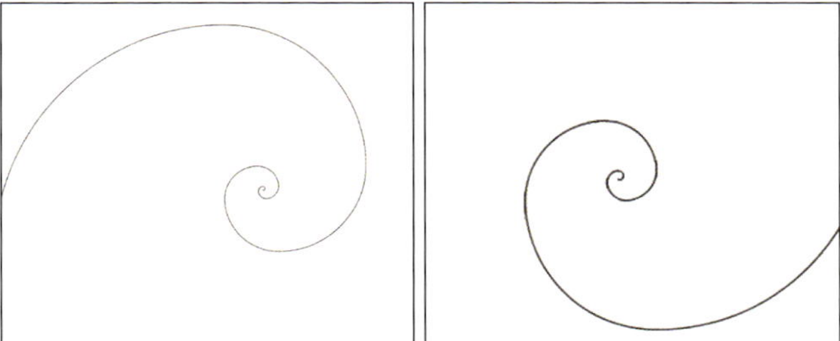

Figure 112 – Fibonacci spirals

Space is like the spring steel wire. If space was stronger, it would be more difficult to twist. Thus if the electron moved to a high-gravity region, more of the electron's mass/energy would have to be exhibited in its electric field. It's as if the Fibonacci spiral is the coil spring from a clockwork toy and you're pulling at the free end. The central area tightens up as you pull, and the finer spiral starts looking like the heavier spiral. Indeed the heavier spiral on the right of figure 112 is just the scaled-up central portion of the finer spiral on the left. That's why it's a scale change, and a change of gauge. If the electron was a knot in your fishing line, it's like pulling it into a tighter knot. The tension on the line increases, the electric field gets stronger, and the fine structure constant takes a higher value.

There's other constants like this. I previously described gravity as a gradient in c, wherein the impedance $Z_0 = \sqrt{(\mu_0/\varepsilon_0)}$ increases, because the strength of space increases. This means c is a running constant, along with ε_0. And if c reduces and Planck's constant of action is momentum multiplied by a distance, that distance gets smaller, so h gets smaller too. *All* these constants are running constants, and one place they run to is the electroweak unification point where symmetry breaking occurs. It isn't like annihilating an electron with a positron, where the resultant gamma photons are like a *twang* from your tight taut fishing line. It's more like when you pull too hard, either on that fishing line or on the Fibonacci spiral. It's more like a *snap*.

And while the running constants are running with increasing gravity, the relative strength of the electric field increases. And since the fine structure constant gives the ratio of the strength of the electromagnetic force as compared to the strong force, a clear conclusion can be drawn:

The fine structure constant is related to kissing numbers, and is a running constant wherein gravity is the result of a gradient in the relative strength of the electromagnetic force and the strong force.

THE STANDARD MODEL

The standard model of particle physics is a quantum field theory that grew out of quantum electrodynamics and quantum chromodynamics in the sixties and seventies. It's generally accepted as being right as far as it goes, but is also accepted as being a work in progress. It covers the particles and the forces, it's made accurate predictions borne out by experiment, but it doesn't cover neutrino oscillations, and it doesn't cover gravity.

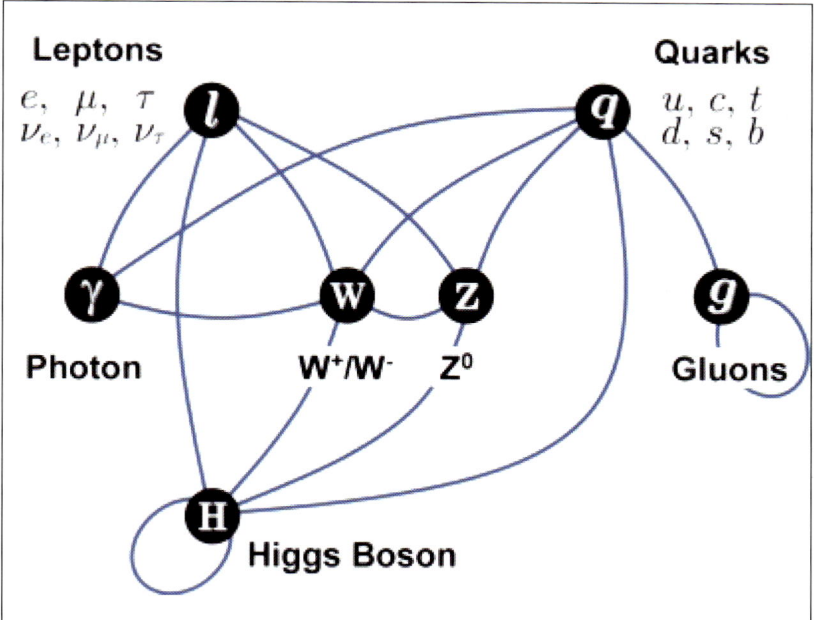

Figure 113 – The standard model

It's incomplete, and as with many mathematical models, there are problems with the interpretation. People tend to think of quarks as billiard-ball particles, unaware that Richard Feynman described them as "partons". The latter delivers a totally different mental picture, wherein quarks are merely component parts. Even Murray Gell-Mann, who chose the name *quark*, thought of a quark as a sub-unit rather than something that could be singularly observed. People tend to forget this, and talk about quark confinement and gluon exchange, taking mathematical abstraction too far and thinking in terms of discrete entities like they think about electrons and photons. In similar vein people talk about observing the W and Z bosons that mediate the weak interaction, forgetting that we've only ever observed the *signature* of the W and Z bosons, not the particles themselves. People gloss over the lifetime of 10^{-25} seconds, they just don't appreciate that these "fundamental" particles just aren't in the same league as the particles we see every day. And nor are they truly fundamental. You can annihilate a proton with an antiproton, and then throw in more particles to annihilate the products, and repeat until all you're left with is photons and neutrinos. All your quarks have gone forever, along with your gluons.

But once you understand the basics, like mass and energy and what particles really are, a new interpretation of the standard model begins to emerge. It involves a retrenchment from quarks back to partons, with messenger particles relegated to accounting units, and *color* and *charm* seen for the mere labels they are. You appreciate that there's no "coupling" with the Higgs field, not when mass is merely a measure of energy that's not moving in aggregate with respect to the observer. As a result the Higgs field becomes a cipher for space, and the Higgs boson is reduced to a tack-on that offered a testable prediction to make a theory more respectable. We just don't need it, not when pair production creates mass, not when the photon is boson enough.

It's space that's fundamental. It's energy that can be neither created nor destroyed. Not the atom-smasher "zoo" of particles that were once thought to be fundamental, and not the lowest-common-denominator quarks that took their place. The more we smashed things together, the more particles we found, because we we were going the wrong way. Because particles are configurations, and quarks are only parts of configurations, and most configurations are not knots. The pions and mesons aren't in the same league as stable particles, they survive for nanoseconds at best, because the short-lived hadrons detected in collider experiments are just temporary fragmentary configurations of twisting turning space. They're "hadronic debris" that lasts for a fleeting moment before decaying, whereupon some of the decay product then decays further. The result is photons and neutrinos, and those familiar stable configurations that we call electrons and protons and positrons and antiprotons. Yes, some configurations may be combinative, but most are more likely to be unstable than not. For example the delta particle has a double charge and must therefore exhibit a double twist, but it only lasts for 10^{-23} seconds. And the theta pentaquark resembles the cinquefoil knot in the middle of the top row overleaf, but not enough, because it only lasts for 10^{-20} seconds.

Figure 114 – Torus knots

All the mystery starts melting away as we understand the geometry of the particles. Like the mystery of missing antimatter. The electron is matter, the positron is antimatter, and the proton is matter too. But given the choice between an electron and a positron, we'd have to say that the *positron* is more like the proton. Because the difference between matter and antimatter is just chirality. If the universe started with a four-square mix of protons, electrons, antiprotons, and positrons, all that's required is a constant creation and destruction along with random distribution to tip us from an unstable situation to one that's stable, where only two opposite chiralities survive. We can then say that matter usually looks like this: ○, antimatter usually looks like this: ⧉, and a hydrogen atom is a combination of both. It's stable because whilst the electron is attracted to the proton like it's attracted to the positron, all it can do is roll around the proton like a ring round a cloverleaf. They're just two travelling warps, circulating spatial distortions that mesh like gears. When the electron changes its energy level, it's changing gear, or changing its path from say a circular s-orbital to a figure of eight p-orbital. Yes, there's another way to make hydrogen, and we call it antihydrogen, but it's like the molecular isomers called *enantiomers*. We don't find left-handed sugar in nature where we live, and we don't puzzle over it.

And nor do we need to puzzle over the unification of the forces. In the chapter on gravity I mentioned the *shear modulus of elasticity*, to do with rigidity. It's different to the *bulk modulus of elasticity*, because it's a lot easier to bend something rather than stretch it out. I also described how Compton scattering changes the direction and energy of a photon like unbending a U-shaped steel bar. Our new picture can be consolidated by considering all the forces in terms of a steel bar.

Properties of the Interactions

The strengths of the interactions (forces) are shown relative to the strength of the electromagnetic force for two u quarks separated by the specified distances.

Property	Gravitational Interaction	Weak Interaction (Electroweak)	Electromagnetic Interaction	Strong Interaction
Acts on:	Mass – Energy	Flavor	Electric Charge	Color Charge
Particles experiencing:	All	Quarks, Leptons	Electrically Charged	Quarks, Gluons
Particles mediating:	Graviton (not yet observed)	W^+ W^- Z^0	γ	Gluons
Strength at 10^{-18} m	10^{-41}	0.8	1	25
Strength at 3×10^{-17} m	10^{-41}	10^{-4}	1	60

Figure 115 – Standard model interactions

The electromagnetic force can be likened to bending the bar, this being achieved with relatively little effort. The electromagnetic field is an all-round bend in space, a twist that alters the path of light to change the motion of electrons and other particles with charge. Move through the twist, and it looks like turn. We call the twist the electric field, and we call the turn the magnetic field, but they're merely two different aspects of the same thing.

The strong force can be likened to stretching the bar. It's more difficult to achieve, because whilst you can easily bend and twist a wire coat-hanger, you can't easily stretch it to make it bigger. The strong force is there in the tensile strength of the trefoil proton that gets harder to stretch as per the bag model. It's also in the electron and the photon, but it just doesn't show. The photon is stretching the bar by 3.86 x 10^{-13} metres, working against the strong force that keeps space together.

The residual strong force that binds neutrons and protons within atoms is something different. It's more like a dance floor where protons are girls with their hands on their hips, and neutrons are boys with arms free to link the girls. The number of girls tells you the element, the number of boys tells you the isotope.

The weak interaction with the neutral current can be likened to planing the bar, or attacking it with a neutrino "grindwheel" along its length. The weak interaction with the charged current can be likened to lathing the bar, or attacking it with a grindwheel around its circumference. So in a sense, the W and Z intermediate vector bosons can be seen as the result of *rubbing*. It's like friction, then something snags and the energy stops, so we've got a pile-up of mass for an instant, followed by decay. Yes the W and Z are actions, and hence we can still call them particles. But they don't persist, they're more like a train crash.

The force of gravity is as if the steel bar is embedded in ghostly transparent rubber, firmly glued to all points on its surface. Bend that bar to stress and stretch the central portion of it, and you're working the rubber too, creating a tension gradient that's the reaction to the action, and the result is gravity. It's a gradient in the strength of space, and in the relative strength of the electromagnetic force and the strong force. The graviton isn't part of the standard model, and never will be, because it's an abstraction. Quantized gravity was always a red herring, based on a misunderstanding of the quantum of quantum mechanics. You can't quantize gravity. Planck's constant isn't the same as the Planck length. A photon doesn't approaches you in steps. It moves towards you smoothly. Gravity is there because light moves slower, because the impedance of space is greater, because space gets *stronger*.

And when c reduces, that common photon amplitude reduces too. So an electron gets smaller, like a tighter knot, and starts looking like a quark. How easy would it be to bend and stretch our steel bar if our steel bar got smaller? It's easy to bend a bar like this —, and that's why the electromagnetic force is weaker than the strong force. But if the bar was heavier-gauge steel and shorter like this - or even like this · , it wouldn't be any easier at all. The fine structure constant is telling us the gauge. It's telling us how strong that bar feels when we try to bend it. It's telling us that we don't unify gravity with electromagnetism, we unify electromagnetism with the strong force, and gravity goes away. The electron gets smaller like the universe gets smaller when we look to the past. That's when electromagnetism was stronger, and gravity was stronger too. Make space smaller, and gravity is stronger. Make space uniform, and gravity has gone. All it ever was, was an exhibition of the unification of the forces, and in doing so, they slip between our fingers, because all they ever were was the geometry of space.

Yes, space is harder than diamond and stronger than steel, because these things and all things are made of it. Because atoms are not 99% empty space. That figure is wrong. Atoms are 100% empty space. There *is* no difference between matter and energy, or energy and space, because they're all one and the same, and the only thing that separates them is geometry. Matter is *made* of energy, and energy and space are the same thing. It's all just space and geometry in action at the heart of it all, and now there's a new way of looking at it all, and now it's all so classical, and so simple, and so very very beautiful:

The weak interaction is mere friction, the residual strong force is mere neutron linkage. The electromagnetic force is caused by twisted space, whilst the strong force is the stretch that keeps space together, and gravity is the reaction and the gradient between the two.

People have struggled for so long to unify gravity and electromagnetism whilst all along the ironic cosmic joke was that gravity itself is an exhibition of the unification of the forces. And the absolute cherry on the top, is that the proton can be modelled as a trefoil knot, and the radiation warning symbol, displayed at all collider sites, is called... *a trefoil.*

Figure 116 – Trefoil warning symbol

THE UNIVERSE

Yes, space is harder than diamond and stronger than steel, because these things and all things are made from it. It's intangible, less substantial than the rarest gas, because it has no substance, because substance is made from space, as matter is made from energy. Because space is distance, and energy is a change of distance, and a change of distance is a distance. It means you just can't have space without energy. And it means the phrase *energy is a volume of stressed space* was telling us something important. Yes, space is like a ghostly elastic solid, but because space is stressed, it behaves like a *compressed* ghostly elastic solid. It's under pressure, so it expands outwards.

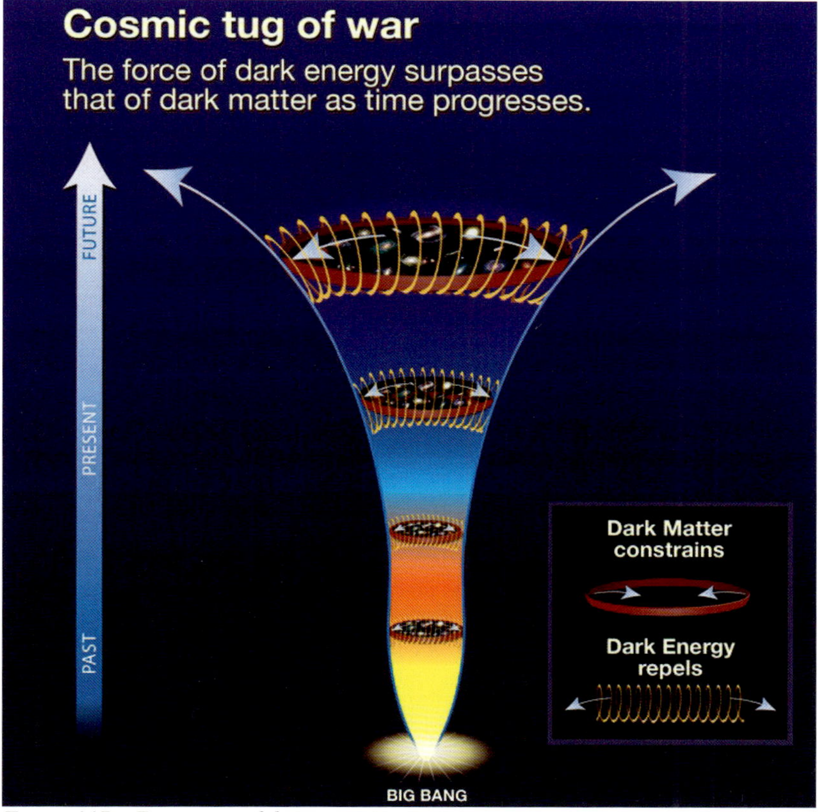

Figure 117 – Expansion of the universe

Yes, some parts are tied in knots and gravitational tension pulls them together, but at the same time there's nothing beyond the universe to hold it all in. And I really do mean nothing. I mean no space, and no distance, and no energy. There's no infinite void of dark empty space beyond the universe. The universe *is* space, and the things that are made from space. There is *no space* beyond the universe, the energy hasn't got there yet, so there is no distance, and there is no *there*.

What this tells us is that the problem with Einstein's cosmological constant can now be seen for what it always was: matter/energy stress cannot be separated from space when matter is made of light, and light is energy, and energy is a volume of stressed space. We must view space itself as energy, and then the mystery of dark energy simply melts away. Space is dark, and it has energy because it's the same stuff as energy, so it *is* dark energy. Space expands because the vacuum of space *is* energy. The universe expands, space expands, distances expand, and the first light expands too, stretching a thousandfold into microwave cosmic background radiation.

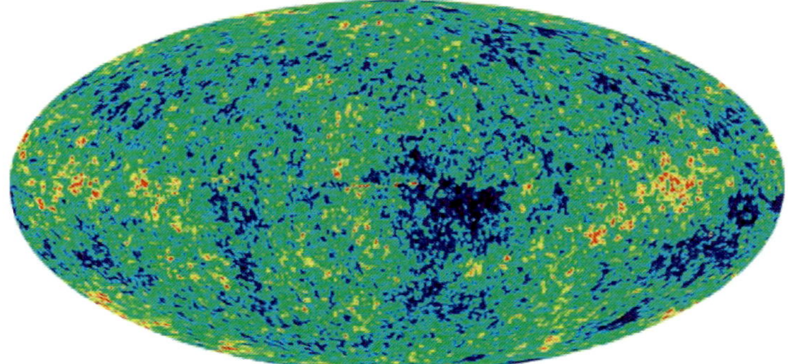

Figure 118 – WMAP map of the universe

It can be hard to get a hold of all this. But just suppose for a minute you could break all the rules of physics and travel faster than light in some given direction. Imagine that the universe isn't expanding, and you're travelling very much faster than light, and you can see the galaxies where they are now, not where they were when the light left them. After a while you might decide that there are more galaxies behind you than ahead, and that you're heading towards the edge of the universe. You keep on going and the effect grows more marked, until eventually there are no galaxies ahead of you. You look behind expecting to see the universe receding. But it isn't.

Because you've reached the edge of the universe, and there's nowhere else to go.

That's what the universe is like. It isn't infinite, it has a size, and it's "flat". Yes, it's unbounded like Einstein said, but it isn't some hypersphere where space curves round on itself and you end up back where you started like some intergalactic globetrotter. It's just a simple three dimensional sphere, but this sphere is expanding, and has been expanding since the big bang some 13.7 billion light years back. You can never reach the edge of it, because you can't catch it up, and beyond it there's no place else, so there is no beyond it. Hence the universe is unbounded, but it isn't infinite.

Even so, as Douglas Adams said in *The Hitchhiker's Guide to the Galaxy*, the universe is really really big. It's bigger than is commonly thought. If the universe had been simply expanding at the speed of light since the big bang, it would have a radius of 13.7 billion light years. If we plotted size against time we'd see a triangle:

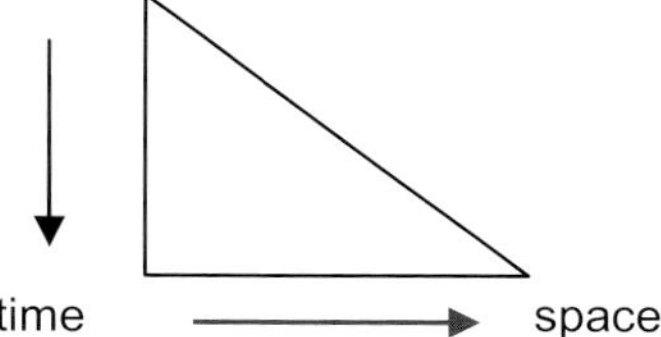

But the universe doesn't simply expand at the speed of light. The observable universe out as far as the "particle horizon" is thought to be ninety three billion light years across[51]. Space is stressed like a compressed ghostly elastic solid, so it expands, then there's more space to expand, and more expanding going on. There's nothing outside to hold it back, so it relaxes and gets bigger. It's getting bigger and bigger faster and faster. As to how big, we have to take an integral, like we take the integral of velocity when calculating kinetic energy. Hence the radius of the universe is more like $\frac{1}{2}x^2$ where x is the age of the universe in billions of years.

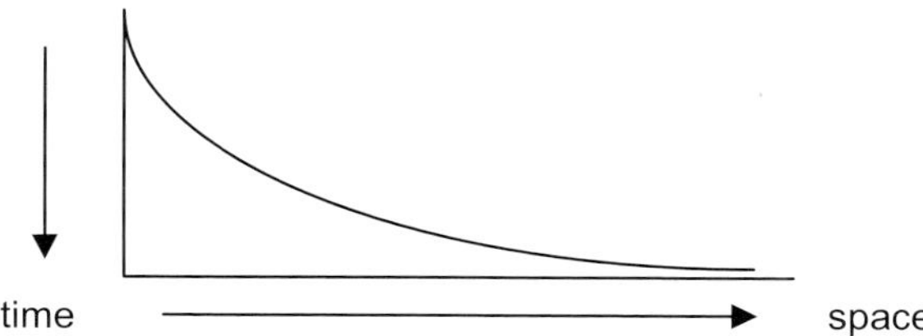

If we take the age of the universe to be 13.7 billion years and square it, the diameter comes out at roughly 188 billion light years. This isn't an exact figure because of inflation and gravity and other things too. The expansion isn't exactly exponential, 156 billion light years is maybe more like it[52]. For all we know it could be a whole lot bigger than that, but it's an interesting illustration. And to illustrate something else interesting, let's take a look at some numbers. Let's look at x^2 values for the series x=1 to 14:

1   4   9   16   25   36   49   64   81   100   121   144   169   196

The increase is increasing, as we can see when we subtract previous terms. Hence the expansion of the universe is accelerating:

3   5   7   9   11   13   15   17   19   21   23   25   27

However the percentage increase is reducing, because as space expands it gets weaker:

300% 125% 77% 56% 44% 36% 30% 26% 23% 21% 19% 17% 16%

So the expansion of the universe is both increasing and decreasing. It sounds counter-intuitive, but it rings true. The universe is running out of steam, the energy of space is spreading out and winding down and weakening, and yet at the same time the expansion of the universe is going exponential. It's a runaway, akin to a ghostly transparent silly-putty elastic that's forever getting slacker, weaker, and relaxing, relaxing, relaxing. And all the while, gravity is losing its grip. When the universe was half the size it is today, roughly five billion years ago, the force of gravity would have been twice as strong. That's because gravity is the tension gradient that balances matter/energy stress, and it's weak because it extends across the universe. So if the universe has doubled in size, the gravity between two distant galaxies must have halved. But the distance between those distant galaxies has also doubled, so the gravity between the galaxies has halved again.

But wait a minute. The space inside the galaxies isn't expanding. Atoms don't expand, because matter is just space with a twist that's tied in a knot. The energy is bound up as mass. And suns are made of collapsing interstellar gas. That interstellar gas hasn't been expanding, it's been collapsing because it's bound by gravity. And galaxies haven't been expanding either, because they're made of collapsing gas too. What it means is that the space between the galaxies has been expanding, but the space inside the galaxies hasn't. And galaxies don't have a surface. There's no cut-off line where space suddenly changes from expanding space to something else. There's a transition, perhaps outside the visible extent of a galaxy, but more likely not.

Figure 119 – Galaxies

What all this means is that space can't be uniform. On the galactic scale, space isn't homogeneous. Where space has been expanding it's weaker. Where it hasn't, it isn't. Space varies in strength depending on whether it's been expanding or not. And gravity is a gradient in the strength of space. This means the further non-uniformity of space that is gravity induced by a mass simply cannot follow the inverse square rule to the letter of the law. That has to be why we see what's called flat galactic rotation curves. When you look at the rotation of the stars and gas in a galaxy, you see something very different to planetary orbits. The outer planets orbit more slowly, and take a lot longer to go round the Sun than the inner planets. It isn't like that for galaxies. Once you get away from the centre of the galaxy, the stars further out don't go slower than the stars closer in. It isn't a smooth "Keplerian" downward curve like it is for planets. Instead the rotation curve goes flat.

People always used to say it was down to a vast halo of dark matter, but Milgrom's MOND[53] was the first attempt to explain it another way. It involves an a_0 cutoff figure of 1.2×10^{-10} m/s^2 where gravity appears to stop diminishing with distance according to the inverse square law, and instead starts diminishing in a linear fashion. This figure of 1.2×10^{-10} m/s^2 is a low value, but that's all it takes to explain the flat galactic rotation curves. Gravity reaches out a little further than expected, so the outer stars orbit faster than expected. It's easy to visualize why. If we've got non-uniform space, the gravity can't follow a neat and tidy inverse square law. It's like the cannonball on the rubber sheet, only the rubber sheet is thinner further away from the cannonball, so the gravity stretches it more. Hence the gravity is more spread out and doesn't follow the expected curve.

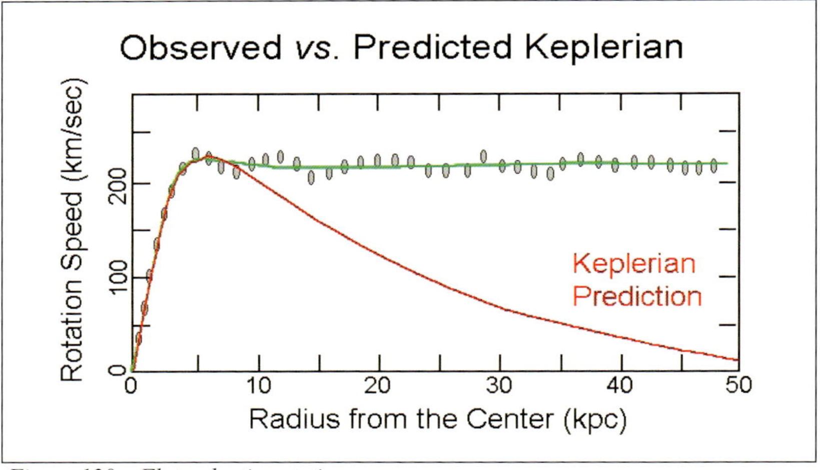

Figure 120 – Flat galactic rotation curve

You can also think in terms of *baseline stress*. Imagine a mass surrounded by empty space. This mass conditions the surrounding space and puts a gravitational gradient into it. The gravitational gradient is a gradient in the strength of space, but if space already has a baseline strength associated with its inherent stress/energy, when you get to a low low level the curve starts stretching out. It's heading for the baseline, not for zero. Then if the baseline stress is dipping down because space is getting weaker away from the centre of the galaxy, the gravity curve follows it down. Look at the upper curve in the figure overleaf. The gravity is stronger than expected at a far-reaching distance. It's stretching out further than the inverse square law says it should.

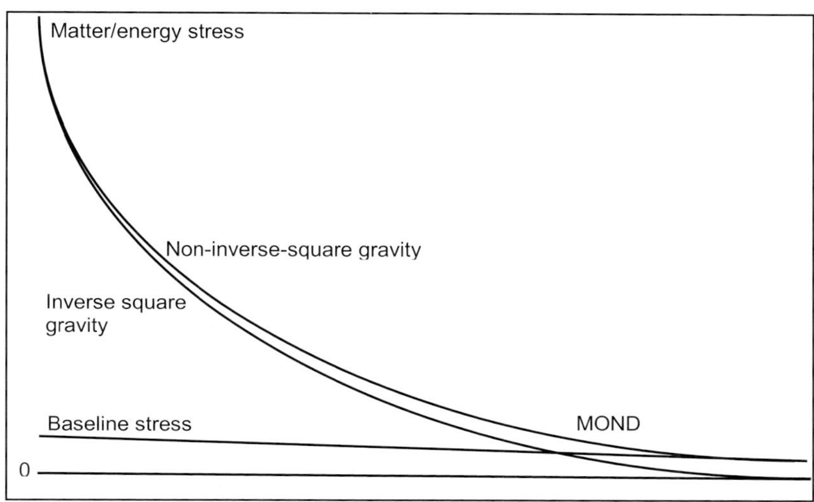

Figure 121 – Inverse square rule

The neat thing about this is that it maybe explains the Pioneer Anomaly[54] too. There's an anomalous Doppler frequency drift equating to an un-expected sunward acceleration of $8.7 * 10^{-10}$ m/s^2. It isn't necessarily a sunward acceleration because what's actually measured is the frequency. The blips should be getting further apart timewise as the Pioneer craft get further and further away from us, and they aren't doing it as much as they should. It sounds like it's to do with the baseline stress of interstellar space. That's higher than the baseline stress of intergalactic space, and in both cases gravity is reaching out a little further than expected. But the jury is still out on Pioneer, because we really ought to see something similar for the outer planets with elliptical orbits, and we don't. And it's still currently out on fly-by anomalies where slingshot spacecraft end up with a slightly unexpected final velocity, as if the planetary gravitational field is not quite symmetrical, like North is uphill. It makes me wonder about non-symmetric metric tensors, and about non-symmetric gravitational theory, but I'll save it for another day.

What seems clear is that the non-uniform expansion of compressed elastic space is a reasonable way of thinking about dark energy, and this seems to do away with the need for huge masses of dark matter. People say gravitational lensing is evidence for dark matter, but actually it isn't. Be it weak or strong, it's evidence of gravitational lensing, that's all, not *evidence for a hypothesis that attempts to explain the evidence*. When we see ordinary lensing it's because the glass is not some uniform sheet, and the image of raisins in a rising cake reminds us that space isn't uniform either. It isn't uniform because the raisins are the galaxies, and they don't increase in size like the cake. Instead of a vast halo of dark matter, all you need is a ring of non-uniformity. Something like a bottle glass window, like the window of the Old Curiosity Shop:

Figure 122 – Bottle-glass space

Yes, there's dark matter out there for sure, be it neutrinos and burnt out stars and gas that doesn't shine, and yes there's ten times as much gas as stars. But to say there's ten times as much dark matter again is pushing it. We have no evidence of WIMPS or higgsinos or other such supersymmetric snarks. But we do have evidence of the thing we call dark energy, because we can see the space, and the red-shift expansion of the universe. The expansion hasn't been homogeneous so space can't be homogeneous. And if we could snap our fingers and take away all the matter in an instant, be it light or dark, all that inhomogeneous space would look like it was riddled with gravitational potentials. It would look like it was full of dark matter.

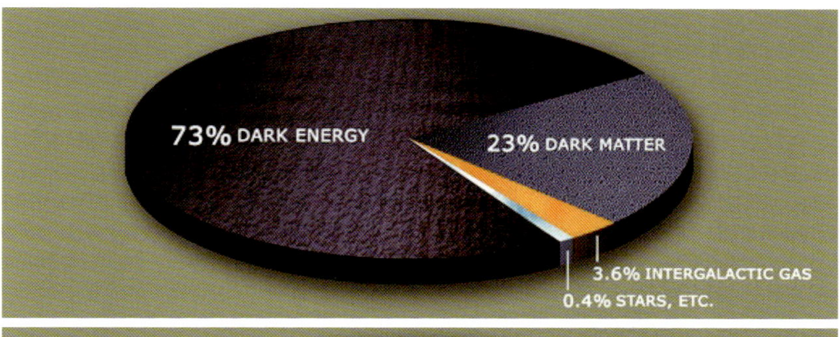

Figure 123 – Dark matter pie

What now seems much more reasonable than dark matter, is that the properties of space vary across space, and over time. We can summarise so:

The Universe expands because space is like a compressed elastic solid, with nothing outside to hold it in. There is no space beyond the universe, there is no distance, there is no there. So the universe is unbounded, but finite and flat. More space expands more, so the expansion is going exponential and gravity is losing its grip. It expands between the galaxies but not within, so space is not homogeneous. It's dark, it's energy, it causes flat galactic rotation curves and gravitational lensing, so we don't need dark matter.

INFLATION COMES FOR FREE

People talk about the heat death of the universe, where everything runs down and the energy is spread so thin that nothing can happen any more. That much is true. They also say the entropy of the universe is increasing. That's true too. But then they say it's always been increasing, and there's the sleight of hand. Do you know what entropy really is? Take the universe and grind it down to the finest powder and sieve it through the finest sieve, and then show me one atom of entropy[55]. You can't. Because entropy is *sameness*. It's an abstraction.

People say that in the early days of the universe, the entropy was very low. But in the early days of the universe, there weren't any days, because a day is just a standing shadow on a rolling ball of rock. There's other things too that just aren't there. People say that when the universe was born it was very very hot. But where was heat before atoms were made? *In the radiation* I hear you say. Ah, photons. Radiation is photons. So where was heat before photons were made? And where was entropy?

Come back with me to the dawn of time. We stand together in a universe less than 10^{-35} seconds old. There are no atoms. There are no photons. There is no motion, there is no heat, there is no time. Everything is the same. Entropy is sameness, so the entropy of this universe is extremely high. There is no matter, because there are no atoms. There is no light, because there are no photons. There is no gravity, because space is uniform. But there is energy. There's lots of energy. And energy is stress and stress is pressure, so there's lots of pressure. That's why you feel something, even within our bubble of artistic license. There's a yawning shift and something gives. It's like a movie when I've just pressed the play button. Things start moving, because the universe is expanding. Now there's motion, now there's ripples, now there's heat, now there's photons. Things aren't the same anymore. Entropy is reducing because the system is expanding, because expansion is motion, so now there's things, and things are moving.

How fast are they moving? You look to your watch, but your watch is running slow. That's because the strength of space is so horrendously high that the vacuum impedance is vast. Those photons are travelling at the speed of light, but the speed of light is a creeping crawl. But still the universe expands, and if we could see through the maelstrom we would see that the universe expands to double its size before your watch even ticks. The initial expansion of the universe isn't particularly rapid. It doesn't really happen especially fast. Space expands like a compressed elastic solid, there's nothing beyond the universe to hold it all in. It does it at its own sweet pace, but because space right now is seriously strong, we and everything within the universe are subject to a titanic time dilation. It's like gravitational time dilation, only there is no gradient so there is no gravity. But it's time dilation in spades, so the expansion seems fast. They call it inflation. And it comes for free.

I always used to wonder why the early universe didn't collapse back in on itself under its own gravity. Now I know why. All you need is a stress ball. You can squeeze it down in your fist then let it go and watch it expand. If you could fill it with lumps and stitch it full of knots it won't expand so fast, but it still expands, it doesn't contract.

Figure 124 – Stress ball

It's the same with gravity. Black holes don't suck in space, they suck in stars. So your knots might flock together, but that stress ball isn't going to squeeze itself down all on its own. Yes as the universe expands the gravity between the galaxies gets weaker, so gravity loses its grip. But it never had any grip on the universe. All it ever had was a grip on the things within the universe. And now it's got even less, because if you waved your magic wand and made a new mass right here right now, the gravitational tension can't extend across the entire universe. It hasn't got time. The universe is getting away from it, like it always was.

Ah, there's all sorts of wonderful stuff in Cosmology. It's my favourite branch of metaphysics, and the science of the very large links so well to the science of the very small. If space is not homogeneous is it fair to say that some parts of the universe are older than others? If space is getting weaker, are there any parts of it that are so weak that they just aren't there any more? Is there really an Axis of Evil and a chirality to the galaxies? If the Earth was formed five billion years ago when intergalactic gravity was twice as strong, was local gravity any stronger and can this be related to jigsaw continental plates? Are gravity waves really the same sort of thing as photons, and does that mean there's other classes of photons with different amplitudes? If photons are transverse waves, how fast can the transverse action happen? Are there any compression waves that run at twice the velocity of the transverse waves, and is there such a thing as a shockwave that runs faster still? Is *that* anything to do with inflation? And what about black holes?

Ah, black holes.

BLACK HOLES

I don't like black holes. Yes, they're fascinating, like monsters always are. But unlike the monsters in the fables, these monsters are real. And they're dangerous. Especially once you understand space, because then you get a new understanding of black holes, and they're more monstrous than ever. But nobody seems to notice, because black holes suck everything in, including a lot of time and energy that I think could be better spent elsewhere. I suppose I'd better explain.

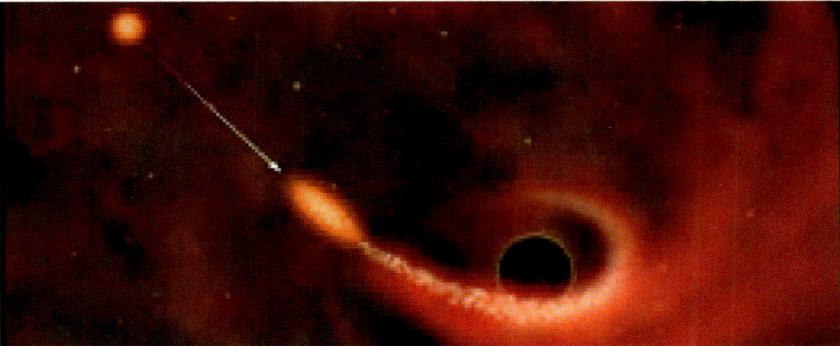

Figure 125 – Giant black hole rips apart star

Benedetti got to grips with gravity as far back as 1553 with the falling balls, something normally attributed to Galileo who was appointed Chair of Mathematics in Pisa in 1589. Galileo got into trouble with the Catholic Church because his telescopes proved that Copernicus was right and the Earth went round the Sun. He was a blasphemer and a heretic, because they didn't have crackpots in those days, and he was also the great granddaddy of relativity. It's called Galilean relativity, wherein you can't detect your uniform motion locally.

Figure 126 – Galilean relativity

In 1687 Newton made the connection between falling apples and the orbit of the moon and worked out the principle of gravity. He was a mathematician and natural philosopher, because they didn't have physicists in those days. OK he was big on Intelligent Design, but things were different then, and what's more important is that he was the father of modern science. He was so far ahead of the game that people gave him grief, and some still run him down as a crank who wasted a whole pile of time on alchemy and the search for the philosopher's stone.

About a hundred years later in 1784 the first person to cotton on to black holes was John Michell. He was a geologist and a rector, but had other interests like inventing the apparatus for measuring the mass of the Earth. He calculated that if a star was five hundred times bigger than the Sun, an object falling towards it from an "infinite" height would end up going as fast as light, so: *"all light emitted by such a body would be made to return towards it by its own proper gravity"*. He called it a *Dark Star,* but nobody paid much attention, because they thought light wasn't influenced by gravity, despite the big corpuscular issue of wave/particle duality at the time. You can find his writings in a book:

Figure 127 – Philosophical Transactions

It's called *Philosophical Transactions* because they didn't have scientists then, they were still called natural philosophers. Even Michael Faraday described himself as a natural philosopher[56] in 1863 but I digress. Fast forward to Einstein and Schwarzschild in 1915 and the idea was resurrected, even though Schwarzschild thought it was just theoretical rather than something real. Fifteen years later Chandrasekhar talked about a limit of 1.44 solar masses above which an inert body would collapse totally. Nine years after that Oppenheimer and Volkoff were on the case, covering neutron stars and upping the bar to 3 solar masses. The collapsed stars were thought of as "frozen stars" where time stopped at the Schwarzschild radius. Nothing much happened until 1958 when Finkelstein introduced the concept of the event horizon as the ultimate one way ticket, then in 1963 Roy Kerr predicted the properties of rotating black holes. In 1967 Jocelyn "no Nobel" Bell Burnell discovered pulsars with a home made radio array, people were suddenly interested, and John Wheeler got the credit for the term "black hole" even though Ann Ewing used the phrase three years previously. The rest is history, and it's been wow-factor ten all the way.

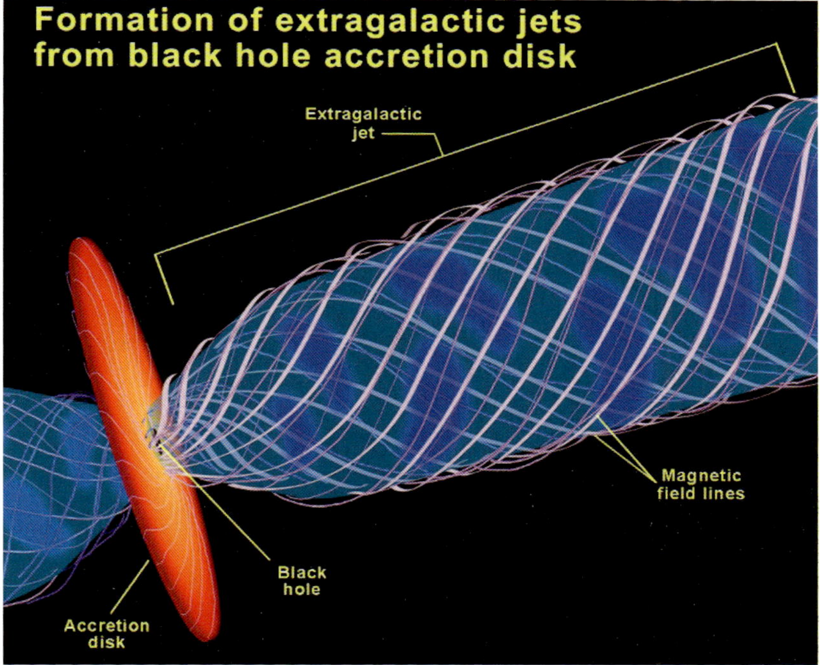

Figure 128 – Black hole with accretion disk and jets

Nowadays we're more confident about black holes, because we've got star formation and stellar evolution pretty much worked out. A star is born when gas and dust form an accretion disk as it collapses under its own gravity. Most of this gas is hydrogen, and the star begins its life fusing hydrogen into helium, releasing energy in the form of gamma photons plus neutrinos. Later it fuses helium into carbon, and fuses more helium into the carbon to make oxygen and a little neon. It's called the Triple Alpha Process.

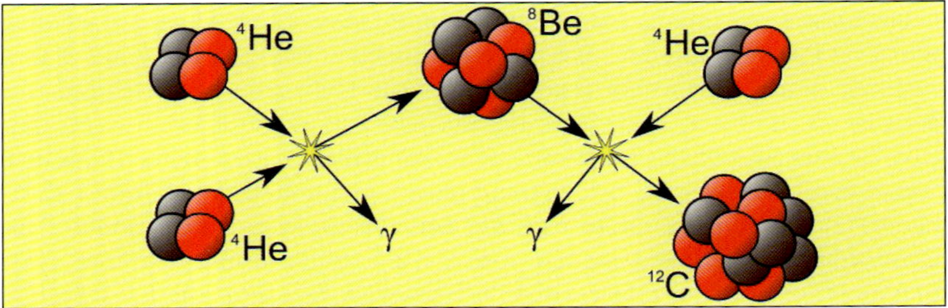

Figure 129 – Triple alpha process

What happens to a star depends mainly on its mass. If there's some heavy elements in the original mix things are a little different, but there's so much hydrogen kicking around that mass is the usual thing. It's all to do with the Main Sequence and the various evolutionary routes that lead from it:

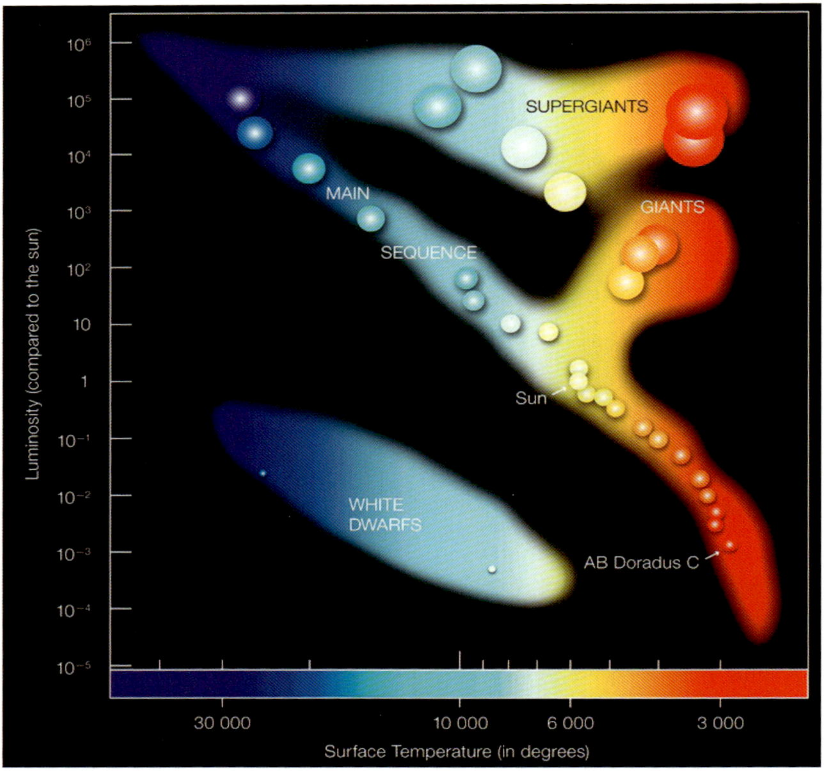

Figure 130 – Main sequence

A small star shines its humble shine for a long long while. Maybe a hundred billion years, far more than the universe has been going to date. A big star burns the candle at both ends, it burns fast, and it puffs out to gigantic proportions. Did you know Betelgeuse is a billion miles wide? You can actually image it through Hubble as a disk rather than a point. It's about twenty times the mass of our Sun, and the core temperature is hot enough to fuse carbon into metals like sodium, and fuse oxygen into silicon. This squeezes out more energy in the form of gamma photons. And they play their part too, in a whole cascade of fusion and fission that produces a liquorice all-sorts of elements including sulphur, chromium, nickel, and iron. All the while there's energy being released and the star shines big and bright. But iron is the end of the line. It *takes* energy to fuse iron, there's no energy to be had from fusing iron. The star coughs and splutters, then runs out of fuel. There's nothing to hold back the gravity. There's only one way things are going to go, and that way is down.

I used to have a mental picture that a star was just gas. Incandescent, like flame, and just as insubstantial. It isn't like that. The average density of the Sun is $1.4g/cm^3$, which means it's denser than me. It's a great deal denser at the core than at the surface, where it's more like gas, only actually it's plasma. And the core of a big star is made of iron, very hot, very compressed, very dense. Imagine holding a stack of bricks a metre high. Now imagine trying to hold up a stack of bricks as high as a house. Imagine trying to hold up a stack of bricks a mile high. Even a brick can't hold up a stack of bricks a mile high. It crumbles into dust. And when you're talking about a stack of bricks a million miles high, even the dust crumbles into dust. It all falls down. The sound of it would be like a million iron skies all falling in at once.

But it doesn't fall straight down at the end. The star has spin, and as it collapses it's like an ice skater pulling in her arms. The spin increases, so much so that the falling material takes on the shape of a disk. There's various things that can happen next, depending on the mass of the star. You might end up with a stellar nebula of gas and dust and a white dwarf in the middle. Or if it started out bigger, a neutron star. Protons and electrons get squeezed together to make neutrons, and they get squeezed together with other neutrons to end up like one giant neutron, spinning furiously. It's only ten miles across, made out of teaspoonfuls of stuff as heavy as a mountain, whirling like a dervish a thousand times a second with spectacular angular momentum and charge galore creating enormous electromagnetic fields that beam out X-rays like a lighthouse.

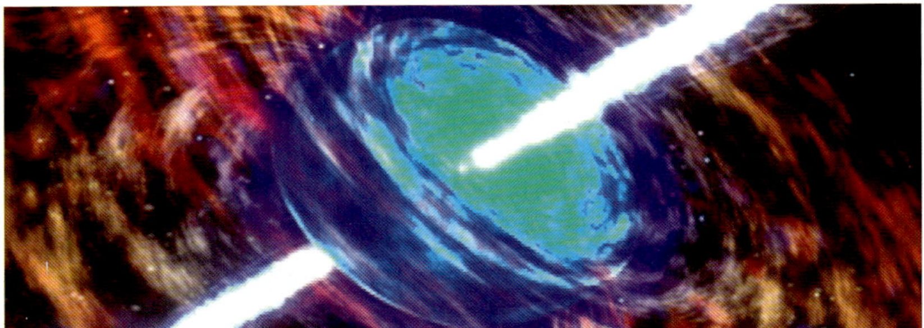

Figure 131 – Neutron star with jet

But of course, if the star was bigger than that, it doesn't stop at a neutron star. Even the neutrons can't hold things up. They get flattened so much they just aren't neutrons any more. What you get is a black hole, and everything falls into it spinning faster and faster like it's going down a plughole. There's still the spectacular angular momentum and the enormous electromagnetic fields, but once everything has fallen down the plughole, it's all gone. Gone forever, utterly *gone*. It's a one way ticket, so it's *gone*, and it ain't coming back.

People don't understand black holes because they don't understand the basic concepts like mass and time and gravity. Remember John Michell's object falling from an "infinite" height? As it approaches the black hole, its relativistic mass increases, and its time dilation increases. At the event horizon it reaches a velocity of c, and everything goes off the scale. The relativistic mass goes infinite, and the time dilation goes infinite too.

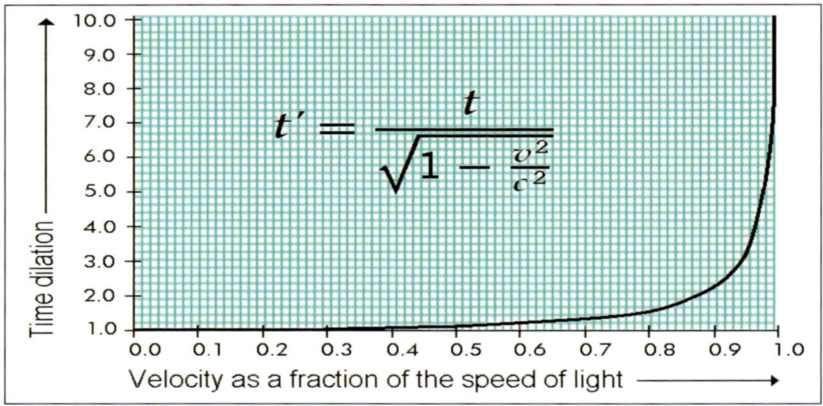

Figure 132 – Infinite time dilation

But we know about mass. An object like an electron is a 511 keV photon travelling in a loop, that's all it ever is, even if it falls past our nose and we see it travelling a helical path like a helical spring. The thing we call relativistic mass equates to one turn round the helix, and the thing we call invariant mass equates to the circular component of this. If the helical spring was stretched out straight, one turn round the helix looks infinite in length, hence the relativistic mass looks infinite.

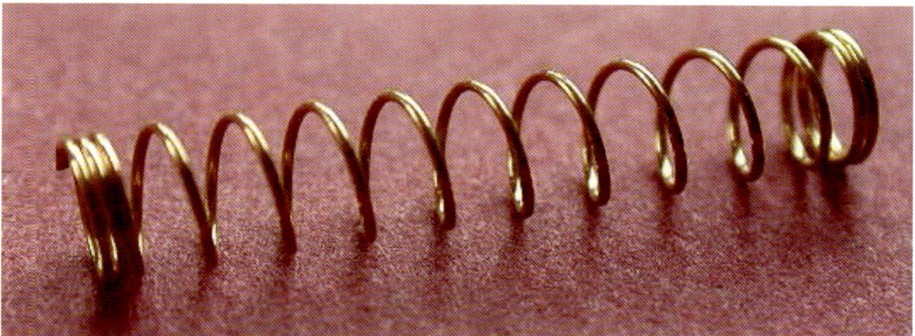

Figure 133 – Helical spring

And we know about time. Time is a relative measure of motion, we can divide the electron's circles into our own to work out the time dilation. But when the spring is stretched out straight we count no electron circles, hence the time dilation looks infinite.

We also know about gravity. We know gravity is the gradient that opposes matter/energy stress, a gradient in space that supplies the kinetic energy of a falling body from that falling body, stolen from $E=mc^2$. Because wherever gravity increases, c decreases. Where gravity is so strong that the time dilation is infinite, c goes to zero. That's the real measure, because there are no infinities in nature. And that means that right now, as we speak, our collapsing star hasn't finished collapsing yet. As far as we're concerned, it takes *forever* to collapse. That means that of all the collapsed stars across the whole wide universe, none have finished collapsing. And they never *ever* will. Hence they're frozen stars. The event horizon is where time ends. And beyond it, *there are no more events*. There is no beyond it. The singularity everybody talks about hasn't happened yet. It's always in the future, over the rainbow, because in truth it exists in a never-never land beyond the end of time. People talk about the "proper time" of an object that falls through the event horizon, but there isn't any. You just can't fall through an event horizon, because it takes forever to do so. You only experience time because of the internal motion of your atoms and electrons and photons. They're all made of energy, a travelling stress of space rippling through space, the same stuff as light. And if c is zero, nothing can travel, nothing can move. Absolutely nothing. Not light, not matter, not energy, nothing. You don't experience any proper time at the event horizon. There are no more events. So there isn't any time. So the proper time isn't *proper* at all.

Now we start to get a different picture of what a black hole really is. It's like solid space, a totally stressed-out place where c is zero, so there can be no motion. It's not just hairless, it's featureless. It isn't a Kerr black hole with spin. If there can be no motion there can *be* no angular momentum, there can *be* no spin. And it isn't a Reissner-Nordstrom black hole with charge. Charge is just twist, the electric twist in space that makes energy go nowhere fast so we call it mass. But all your electromagnetic charge has been chopped off at the event horizon along with the light, leaving your magnetic flux flailing and your charge gone forever. It's the same with information. It's gone forever, melted like plastic letters thrown into a cauldron. All that's left is the stress, the energy frozen into what we call mass exerting its gravitational tension in the surrounding universe. Anything falling into a black hole just adds to the total stress, hammered flat by the total radial length contraction whilst making the black hole bigger.

And it gets worse. Because bigger isn't the right word. Because light defines our time, and our distance, and when c is zero, there is no time and there is no distance. We measure time and distance using light, and we simply cannot measure these things inside a black hole because light will not propagate when c is zero. So a black hole is like Doctor Who's *Tardis* in reverse. It's a *no-time* machine. It's smaller on the inside than it is on the outside. On the outside it might be a supermassive 12.5 light hours across. But on the inside it's zero light hours across.

Put your hand over the picture on the next page. Don't look at it just yet. It's supposed to show you gravitational lensing, the refraction associated with gravity. I want you to think of a balloon. Think of a party balloon. Think of a black party balloon, made of rubber, maybe a little translucent, kind of see-through, with a greyish tinge. Only this is a special party balloon, because I've painted a starscape all over it. This party balloon is an analogy of our ghostly elastic space. OK, are you ready? Take a deep breath. This is visceral, because this is where you get to understand what a black hole really is. Now take a look at the picture:

Figure 134 – Simulated image of a stellar black hole

Can you see it? The balloon has a *hole* in it. The black hole really is a hole. It's a hole in space. The only properties this hole exhibits is size and mass, and the mass is just stress, the stress that counters tension and causes gravity. And just like an ordinary hole it only has these properties by virtue of what it's a hole in. If the hole got bigger there would be less surrounding space, there would be less "balloon". If it got as big as one whole hemisphere maybe the stress sweeps over the curve and turns everything inside out. If it just got bigger and bigger in the end there would be no balloon at all, no stress and no tension, no space and no time, and no hole either. Everything would be gone. Because a black hole is *something and nothing*. It's a gravastar[57], a void in the fabric of space and time. Because it's solid space and no space too. Because a hole in nothing is something, and it really is a hole.

The mathematics of general relativity is said to blow up at the central singularity. But if there is no central singularity, the maths doesn't blow up. We need to take account of a different view of time, and a relocation of the singularity to the event horizon. But it's not an infinity, because it isn't really infinite time dilation, it's a *c equals zero* at the Schwarzschild radius. It's the end of events, the end of motion, the end of time, and the end of space. There is no inside, there's only an outside, and that's why it's like a surface. It's a mathematical surface, because there's no real surface, but that's why the entropy is proportional to the surface area. It's the holographic principle in action, it's like the universe inside out.

Figure 135 – Hyperbolic

It's maximum entropy and maximum sameness and it's just a hole, a nothing thing, a bubble, a shell, with nothing inside it because there is no inside it. And at the same time it's a something because everything is relative and a hole is a something:

Black holes are frozen stars, where gravity is so strong because space is so strong that c is zero, so light can't move, nothing can move, and nothing can fall through the event horizon. It's the end of time, the end of events, and there is no singularity and never will be. So there's an outside but no inside, it's solid space but no space too, a something and nothing, a bubble, a shell, a hole in space, like the universe inside out.

This is the stuff from which the world is made. It's something and nothing, and it's pure energy too, the sort that at last you can hold in the palm of your hand. But I wouldn't advise it. Because grasping this pure energy would be finding out about something and nothing the hard way. It would be like putting your hand into a meat grinder, only far far worse. A crackle and a crunch, and then you're gone. You're gone forever, and you ain't coming back.

Yes, black holes are interesting all right. Especially when you see them in a new light. For example, how exactly do black holes undergo gravitational attraction? Normal matter is subject to a gradient in c across the local frame that effects a refraction. But everything about a black hole is frozen to nothing, so there's no refraction to make it move towards another mass. It's rock bottom on the tension gradient and the only way to shift it is through some kind of surface effect. I know how to move a soap bubble, but I don't know how to move a hole in my rubber sheet. So I start looking at unusual orbits and off-centre galactic cores in a new light. Just as I look at Pace VanDevender's Irish bog[58] in a new light. I also look at optical black holes in a new light. So I think of Ronald L Mallett trying to build a time machine, and I look at that in a new light too[59]. I think the best he can hope for is a *no-time* machine, a *stasis box*. You put something in the box, and nothing happens, no time is experienced, so it's like the ultimate re-frigerator. But there's something about it that makes me feel a little uncomfortable. The world is painted in light, and so is the canvas. So I can't see the difference between an optical black hole and a real black hole. Could this thing be dangerous? Is there any way it could get bigger of its own accord?

THE BIG BANG

Like most people, I like to think about how it all began. I know that in a simple allegorical way, people talk about space like it's a gas, while matter is akin to a liquid, and a black hole can be likened to a solid. I can imagine a sublime phase change where a solid is instantly converted into an expanding ball of compressed gas, some of which condenses into droplets of liquid. And yes, I can understand the separation of the forces via the simple geometrical picture, and I can also understand how gravity pulls some things together. But all that *first three minutes* stuff doesn't really satisfy. I start with the expanding universe, hit the rewind button, and watch it wind back all the way to the beginning. But then my screen goes blank, and then it disappears altogether.

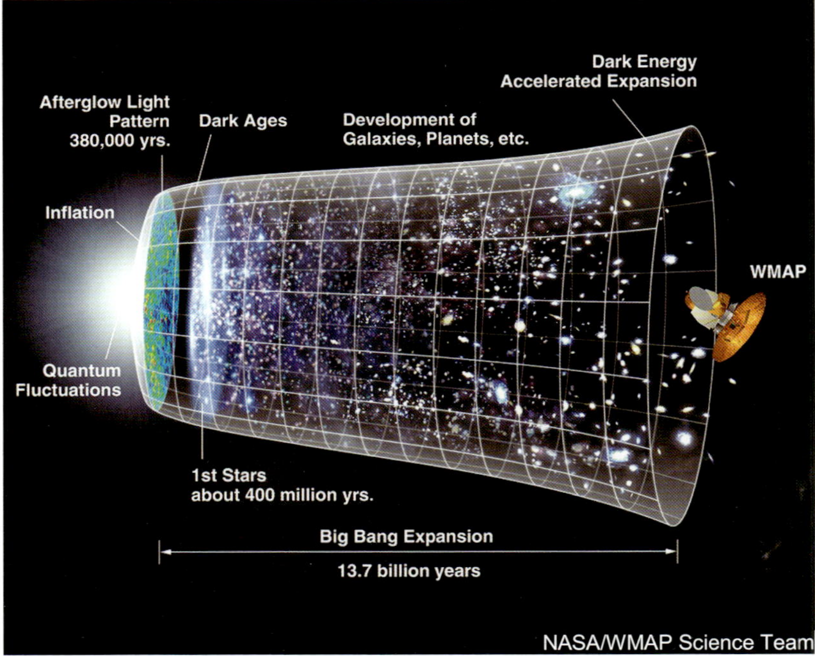

Figure 136 – Time line of the Universe

The problem is that if all the matter and energy of the universe came together, there would be no stress, no tension, no light, no energy, no distance, no space. And the Universe would look like this: Get the picture?

There is no picture. It would be nothing. Nothing at all. The universe is only *manifest* because of the stress and tension. No stress and tension means no distance and no energy and no gravity. And that means no universe, and nothing to create it from. That's when I hit the buffers. I don't know how to get something from nothing. I just can't conceive of nothing, because nothing doesn't exist. So I don't know how to make a big bang. I'm stumped.

The only straw I can clutch seems to be black holes. They're something and nothing, something to chew on. I know how to make a black hole, even though I don't know how to make a universe, not out of nothing. But if I knew a way to get things out of a black hole, that would make it a white hole. *That* would make it something like the big bang. And there is supposed to be a way to get something out of a black hole. It's called Hawking radiation.

Figure 137 – Hawking radiation warning sign

In 1974 Stephen Hawking "discovered" Hawking radiation, where hypothetical micro black holes evaporate due to hypothetical virtual particles being created from the vacuum and getting separated. One carries negative mass/energy into the black hole to balance the books whilst the other escapes to carry positive mass/energy out into the universe.

I don't buy it. Negative mass would involve runaway repulsion. It's a negative carpet, we've never seen it. And we've never seen virtual particles either. They're *virtual*, they break the rules because they can propagate faster than light, and they create infinities which have to be swept under the carpet by renormalization. They're just an accounting convention for mathematical point particles. Real particles are virtual enough, even fermions that are merely twisted bosons going nowhere fast. I certainly don't buy it for fermions like the electron and the positron because both are made out of positive energy. There's no negative mass/energy here. We create electrons and positrons via pair production, and we annihilate them in PET scans. It's real. It's science. Not a hypothesis built on abstraction.

I still don't buy it for bosons like photons, because photons are positive energy too. Yes, the photon is its own antiparticle, you can combine two out-of-phase photons and be left with no photons, but the energy is now in space. It's what people call zero-point energy, detected by the Casimir effect. People talk about transient variations of the zero-point field, where small regions are temporarily above and below the local ground state, like wavelets on the surface of the sea. It's reasonable enough. But to then think of these quantum fluctuations as virtual particles selectively transporting negative energy is going too far. As far as I'm concerned space and energy are the same thing. I see no negative space. The zero-point energy of intergalactic space might be below our local ground state, but that doesn't make it negative energy. The only "negative energy" I can see is the gravitational tension that opposes matter/energy stress. It's the other side of the coin, the two can not be separated. And that's enough for me to say that if somebody did manage to create a micro black hole, I'm not convinced that Hawking radiation would come to our rescue.

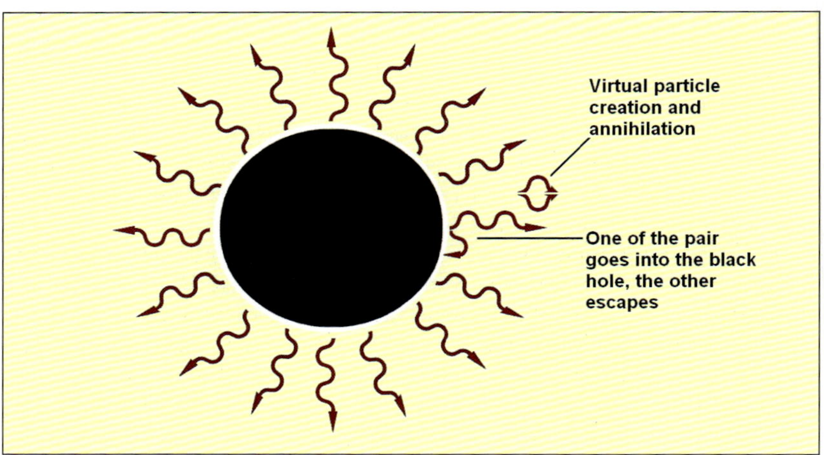

Figure 138 – Hawking radiation

Let's assume particles can be created from the vacuum, and this involves photons, the most common and the most fundamental particles we know. Photons move energy. That's what they are, that's what they do. If the energy of space near a micro black hole is somehow separated into two out-of-phase photons, the micro black hole might swallow one and lose the other. One photon would escape into the universe, and we'd see what looked like Hawking radiation. But the black hole wouldn't be evaporating. Worse, the black hole could swallow both photons. The stress and tension balance, they still add up to zero. The stress that is energy is balanced by the gravitational tension. So the micro black hole never has to repay any virtual particle energy debt, because there is no debt. This kiddie is his own banker, and he keeps his own time. What it means is that Hawking radiation could be back to front. A micro black hole could *grow* due to QM effects, increasing its mass and gravity, apparently creating both matter/energy stress and gravitational tension out of nothing. It would grow bigger like a Hokey Cokey dance circle, consuming our space. It sounds like a runaway, and I don't like it. Especially when I read Neil Cornish, an astrophysicist at Montana State University, saying this:

Indeed, the fluctuations we see in the CMB are thought to be generated by a process that is closely analogous to Hawking radiation from black holes.[60]

I don't like the idea that the fluctuations in the cosmic background radiation look like Hawking radiation, particularly with what I hear coming from the Large Hadron Collider. Yes I know cosmic rays have been measured at 10^{20} electron volts, but when I hear bona-fide physicists at CERN talking about *energies not seen since the big bang* and *creating micro black holes*, I feel uncomfortable. The risk assessment cuts no ice, because it isn't independent. They say there's no risk, but they're hoping for the unexpected, and it just doesn't square. People are experimenting with things they don't understand in order to understand. And they're so keen to do so, they just won't stop to look and listen and understand just how simple it all is. Instead they're wrapped up in a groupthink that promotes the mystery of mass when there is no mystery. They spin the yarn about the fabled Higgs boson and unlocking the secrets of the universe, when the answers are already there. Oh I don't think the Large Hadron Collider will create any black holes. I don't believe in those extra dimensions that would help things along. Nor do I believe in strangelets. I don't believe the "God particle" will be found either. Instead I take the view that we'll obtain unspectacular information. It'll be nothing to write home about. It'll be a damp squib. It'll be CERN's nightmare scenario, ten billion dollars to prove nothing worthwhile, when all the while there were better things to spend the money on. Things like fusion and photocells, and high-density batteries. And other things too, like space missions and gravity, and Jodrell Bank.

What I'm really uncomfortable about is the aftermath, because I know it's a no-win situation. Once the hoopla dies down, the guys at the LHC will be in for the long haul, and meanwhile science will advance. And then I think some people will sit down and think about *there's no conceivable risk* and *we're hoping for the unexpected*. I think they'll start asking awkward questions, this time for real: Are the LHC guys just playing with knots as they chase the prize? Are they playing with fire? Or are they playing with dynamite?

Figure 139 – Alfred Nobel

I think they'll ask those questions because there was no beginning of time, because the time is always now. We can't create something from nothing, because nothing doesn't exist. How can a prime mover make the first motion when there's nothing there to move? People used to think the Earth was flat, supported by four elephants, standing on a turtle. But they couldn't say what supported the turtle, so the answer was a non-answer, it was *turtles all the way down*. Religious folk say God created the universe, but that's no answer too, because who created God? A quantum fluctuation is more of the same, and so is the primordial atom. So the only place left for me is that the universe is eternal after all, despite what we see. It means the big bang wasn't the beginning, because there was no beginning. And I ask myself this: *Did the bible just pop into existence all on its own? Did God create it? Or did people create it?* On balance of possibilities, I'll go for people. Now replace the word *bible* with *big bang*. If one day there's a blinding flash and it's game over, you'll know what I mean. But there again, I suppose you won't. Nobody would ever know.

And there's the rub. Once people understand how things work, they'll understand things like high energy physics and the psychology of belief. They'll understand no-win situations, why nobody stands up to be counted, and they'll understand risk. Then they'll ask those questions, and one thing they'll find out is that some people took the view that there was never any risk, because if it all went wrong nobody would ever know. Then people will say *What were you thinking?* Not just about high energy physics, but about other things too, like time travel and branes and parallel worlds. That's what I feel uncomfortable about. That's my nightmare scenario. That when it's all done and dusted, people will look back at the absurdity of it all, and science will be the poorer for it.

RELATIVITY+

As it happens, I'm not that keen on the *Theory of Everything*. It's too sweeping, suggesting that physics is all done and dusted and there's nothing left to do. It isn't like that. What I've been telling you about is better described as a *Unified Field Theory*. Or better still a *Unified Model*, because the fields are mere geometry, and it's only a conceptual model. It's what you call a *qualitative* model, and it needs to be a *quantitative* mathematical model before it can be officially considered to be a theory.

There's a lot more to do before it gets there, but it's a good start because it delivers understanding like never before. You understand that you can't understand axiomatic terms like E=energy and m=mass using mathematics. You realise that something like t=time is treated as a self-evident truth that you don't even need to look at, and as a result Minkowski spacetime is taken for granted. So much so, that many people think that any challenge to it is a heresy, tantamount to undermining relativity itself. It isn't like that. Seeing time for what it is doesn't topple relativity, it unleashes it. People don't realise this, and they don't realise that you have to go top-down to get to the bottom of things. Mathematics is vital for calculation and prediction, but it isn't so good for conceptual understanding. That's why after all these years people still don't understand simple things like energy and mass and gravity and particles. You can't understand these things using mathematics, and that's what Gödel's Incompleteness Theorem was always saying. But you *can* understand them with logic, and analogy, and a cold rational eye that looks at what's real.

When you look back you see what people were groping for. Think of a cube, then apply a distance variation. The cube is three dimensional, denoted by x y and z. It can extend in the x direction, or the y direction, or the z direction. Or in the xy direction, the xz direction, or the yz direction. Or in the xyz direction. Or in no direction, because you're free to pause.

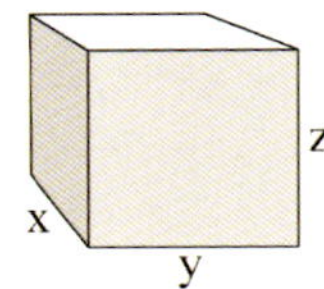

That's eight degrees of freedom. Not dimensions. Wilhelm Killing's E8 root system consists of 240 vectors in an 8-dimensional phase space, but that's a mathematical space. What E8 is really telling us is about dynamical geometry in three dimensions. It's Murray Gell-Mann's eight-fold way, as sculpted by Helaman Ferguson:

Figure 140 – Helaman Ferguson sculptures

It's all just geometry in three dimensions of space, where space extends in different ways according to the least-action principle that belies its elastic nature. Once you realise this, you understand why fields are geometry, why you can't quantize gravity, that there are no point particles cannoning around empty space. And you understand that there are no strings, because whilst better than zero-dimensional points and undoubtedly elastic in nature, one dimensional strings don't go far enough. They tell you nothing at all about what they really are, and nor do branes, which go too far.

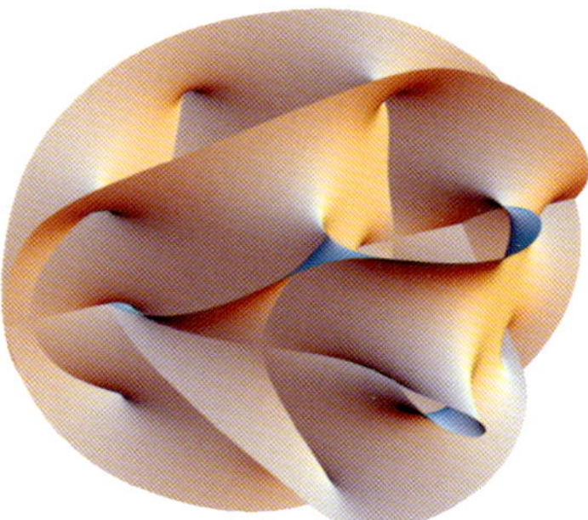

Figure 141 – Calabi-Yau manifold

You also understand why all the quantum magick evaporates in a trice. There is no multiverse. There is no collapse of the wave function just because you observed it. There are no pilot waves steering particles through slits. The probability doesn't indicate where the particle will be found, it indicates where the particle *is*, and it isn't in one place. It's an extended entity, a distortion. The wave function *is* the particle, because it's a wave of space in space, be it a transverse wave, a rolling wave, or a soliton wave tied in a knot. Once you understand all this, a picture emerges, one that's so coherent and so crystal clear that you wonder why you haven't seen it before:

In barest essence energy is a volume of stressed space.

Mass is a measure of the amount of energy that is not moving in aggregate with respect to the observer.

Charge is curl, charge is twist. The electromagnetic field is a region of twisted space, and if we move through it we perceive a turning action which we then identify as a magnetic field.

Time exists like heat exists, being an emergent property of motion. It's a cumulative measure of motion used in the relative measure of motion compared to the motion of light, and the only motion is through space. So time has no length, time doesn't flow and we don't travel through it.

Gravity is an extended tension gradient opposing matter/energy stress, wherein the speed of light varies resulting in gravitational time dilation and attraction through refraction.

Space is a one-trick pony, and the only trick is distance. The photon is energy, and is a rippling change of distance that makes for more volume. A change of distance is a distance, so space and energy are the same thing.

The photon is a wave of distance variation. We can tie it in knots to make particles with mass and charge. The electron is a trivial knot, the positron is its mirror image. So matter is just space with a twist that's tied in a knot.

The common photon amplitude is a spatial extension of 3.86×10^{-13} metres, and is the quantum of quantum mechanics. The wave function doesn't describe where a point particle can be found, it describes where the extension is.

The electron is a trivial knot, a turn and a twist. The proton is a trefoil knot, three turns and a twist. The neutron is a proton plus a twist and two turns. The neutrino is a turn, a mere running loop, and muon and tau neutrinos have more loops, as do the muon and tau themselves. The antiparticles are mirror-image knots that go the other way, and the unstable particles are not knots, so they always come undone.

The fine structure constant is related to kissing numbers, and is a running constant wherein gravity is the result of a gradient in the relative strength of the electromagnetic force and the strong force.

The weak interaction is mere friction, the residual strong force is mere neutron linkage. The electromagnetic force is caused by twisted space, whilst the strong force is the stretch that keeps space together, and gravity is the reaction and the gradient between the two.

The Universe expands because space is like a compressed elastic solid, with nothing outside to hold it in. There is no space beyond the universe, there is no distance, there is no there. So the universe is unbounded, but finite and flat. More space expands more, so the expansion is going exponential and gravity is losing its grip. It expands between the galaxies but not within, so space is not homogeneous. It's dark, it's energy, it causes flat galactic rotation curves and gravitational lensing, so we don't need dark matter.

Black holes are frozen stars, where gravity is so strong because space is so strong that c is zero, so light can't move, nothing can move, and nothing can fall through the event horizon. It's the end of time, the end of events, and there is no singularity and never will be. So there's an outside but no inside, it's solid space but no space too, a something and nothing, a bubble, a shell, a hole in space, like the universe inside out.

I hope that you have now been able to grasp these basic concepts, and have formed a mental model of how the universe works. It won't be perfect, and it isn't complete, but I like to think it's enough to give you a better handle on things, and give you that intuitive knowledge that you may have previously lacked. It's enough for you to explain it to your grandmother, like Einstein said you should. And to stop fooling yourself, like Feynman said you should. Then maybe you can make predictions, and formulate experiments, and do the maths. Then maybe you can take it further.

For myself, I've always held Einstein in the highest regard. Special relativity and the general relativity that subsumed it was always about electrodynamics, and energy and mass and the speed of light. Once you see the crucial role of motion, you're free to look at these concepts with fresh eyes. And every step of the way you see relativity, in the energy, the mass, and the charge, in the time, the gravity, the space, and the particles. It recurs time and time again. That's why I'm giving this model the name RELATIVITY+. Perhaps Einstein had a better name when he was working alone in his twilight years. I'm not sure. But I am sure that light is tied up as matter. We are made of light. That's what $E=mc^2$ always meant. Light is energy, and energy is the stuff of space, so matter is just space with a twist that's tied in a knot. The particles are not points, they're not billiard balls, they're not strings. They have no surface, they're distortions, some running free, some not. And the things we detect them with are distortions too. That's why we can never pin them down. That's why they're uncertain, and why the wave function is the particle. It really is a wave, of space in space. That's why there's no magic, and nobody plays dice. No wonder Einstein considered quantum theory to be an imperfect description of reality. No wonder he was trying to come up with a unified field theory where the concept of field was no longer appropriate.

"Of course it would be a great advance if we could succeed in comprehending the gravitational field and the electromagnetic field together as one unified conformation. Then for the first time the epoch of theoretical physics founded by Faraday and Maxwell would reach a satisfactory conclusion. The contrast between ether and matter would fade away, and, through the general theory of relativity, the whole of physics would become a complete system of thought, like geometry, kinematics, and the theory of gravitation."

It was always all about geometry, and 3+1 dimensions instead of 4. Once you take the time out of spacetime, you see that it's the electromagnetic field that's curved space, not gravity that's curved spacetime. In unifying the forces, you come to see that they're all pseudo forces, that the fields aren't fundamental at all. Instead they slip through your fingers and you let them go as you grasp the geometry instead. It's Einstein's pure marble geometry, it's Einstein's dream.

THE SAME ELEPHANT

The history of science has been incremental, a clue here, an idea there. Yes, sometimes somebody joins the dots and we get a breakthrough, but more often than not we inch forward using the input from many many contributors. Science is a collective effort, methinks a far richer tapestry than most people realise. Yes there are heroes, but there are unsung heroes too. For every hero that catches the public eye there's a dozen who don't. And for every one of them, there's a hundred, no a thousand men and women who have made vital contributions that are forgotten, lost in the mists of time, swept under the carpet by the airbrush of history.

Figure 142 – Tapestry of the Centuries

So it's time I gave you some background material. It isn't just Einstein and Dirac and De Broglie and Schrodinger and Maxwell and Faraday and Newton who thought along these lines. There's lots of other people, more than you realise, and probably more than I do too. For all I know, everything I've said has already been said, ten times over in dozens of books and papers and articles that I haven't seen. But there are books and articles and papers that I *have* seen, and I want to tell you about them, along with the people involved. Yes, much of the material is straightforward, factual and uncontentious. Some is not. I want to tell you about it all, and I want to set it in my own personal context because there's something else I want to say: I owe a debt of gratitude and acknowledgement to the various individuals and organisations that I've drawn upon to produce this work. They are the unsung heroes.

I go by the name of *Farsight*. That's my username on the internet. It's the meeting place, the nexus, more important than most people know. Perhaps in a decade we'll know for sure, and we'll shake our heads in sorrow that people like Einstein didn't have the internet. Oh, I was interested in physics long before we had the internet. I've been interested since I was a kid, when I read my mother's school-prize science book, the one with Kipling's serving men on the inside cover:

I keep six honest serving-men. (They taught me all I knew).
Their names are What and Why and When. And How and Where and Who.

But I was on slow burn all my life until the internet brought it to life. The internet turned it from an interesting hobby into something that lit my fire. It stopped being a one way street where I learned textbook teaching. It became a two way street, where I started thinking for myself. It started in 2005 when I came across an article called *Energy Misdefined in Physics* by Gary Novak, see http://nov55.com/ener.html. It concerns kinetic energy and momentum, and whilst I didn't know enough at the time to understand whether he was right or wrong, he makes some interesting points, and I was interested. It got me thinking. This reawakened my interest in physics more than anything else.

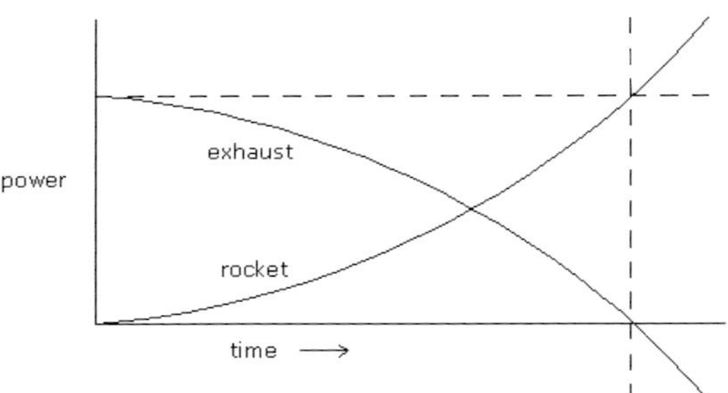

Figure 143 – Rocket power

I guess I entered the fray at that point. I enjoyed a great deal of interesting discussion and learned a great deal. Fast forward a year to the middle of 2006, the Maths tower had been demolished, and my teenage son and daughter told me they were dropping all of their science subjects. That's when I found out that the experiments were pendulums instead of Van Der Graaf generators. That's when I found out that A-level physics had declined by 57% in twenty years. And that's when I decided that I was going to do my bit. So I made an effort to explain the simple stuff and gave homework help on physics forums. But the naive questions caused me problems. I couldn't answer them, and when I tried, I found myself stepping out of line. People didn't like me trying, wouldn't believe what I said, and wouldn't even consider the subject properly.

That's where *THE PSYCHOLOGY OF BELIEF* came from. I was astonished to find that sometimes people won't apply logic, and reject anybody else's attempt to do so. I'd seen this sort of thing before when talking to Young Earth Creationists, and I was seeing more of the same from scientists. When people have some form of belief, there's just no talking to them. OK I didn't write it up until later, and I'd finished the first cut before I'd ever heard of Thomas Kuhn. But I'd like to say thanks to him and all the other people who've written about this sort of thing. I can't remember the details of what I've previously read and when. But something sank in somewhere along the line. Until you clock it you don't realise just how much utter conviction there is. It's everywhere you turn, and it can be pig headed, close-minded, and downright nasty. I remember one guy was particularly unpleasant about the speed of light, whereafter I looked up at my bookshelf and pulled down a couple of books. I weighed those books in my hand and thought *time is the key*. I resolved to get it down properly. I'd use my analytical skills to tackle those naive questions and write an internet essay. I'd apply my experience to make it easy reading, understandable to the layman, and yet so logical that nobody could object.

And so *TIME EXPLAINED* was born. That was my first internet essay, and actually it's nothing new. The concept of time therein has been around for quite a while. The couple of books I mentioned were *The End of Time* by Julian Barbour and *About Time* by Paul Davies. They date from 1999 and 1995 respectively, and I originally read them in 2001. Neither of them hit home at the time. I wasn't paying too much attention, I was sleepwalking. But thank you Julian and Paul because something lodged in my mind, along the lines of *time is change*. It isn't a brand new idea, because it goes back to Aristotle. But I could put a fresh slant on it, using user-friendly language and a tight delivery that would be crystal clear. By October 2006 I'd written the first version of *TIME EXPLAINED* and posted it on a forum. The optical illusions were important even then, because I was making the point that people take too far much for granted. Hence I'm grateful to "echalk" *Online Resources For Teachers*, see http://www.echalk.co.uk/amusements/Optical Illusions/illusions.htm, and to R Beau Lotto and Edward H Adelson who created the particular illusions I used. OK that first version wasn't much to write home about, but I asked for feedback to help improve it, and got it. It was interesting and informative. I learned about Presentism. A fellow forummer called mganderson flagged up *A Hole at the Heart of Physics; A Matter of Time; Special Editions*; by George Musser in the August 2002 issue of Scientific American. THoR gave vital encouragement, as did amrit, real name Amrit Srecko Šorli, who was ahead of the game but language was a barrier. Then on 29th October 2006 yquantum recommended a book called *A World Without Time: the forgotten legacy of Gödel and Einstein* by Palle Yourgrau. It's historical and philosophical, heavy going at times, but pure gold dust. I was amazed to discover that Einstein thought of time this way too. Not in 1905, but in 1949, when he was at Princeton with Gödel.

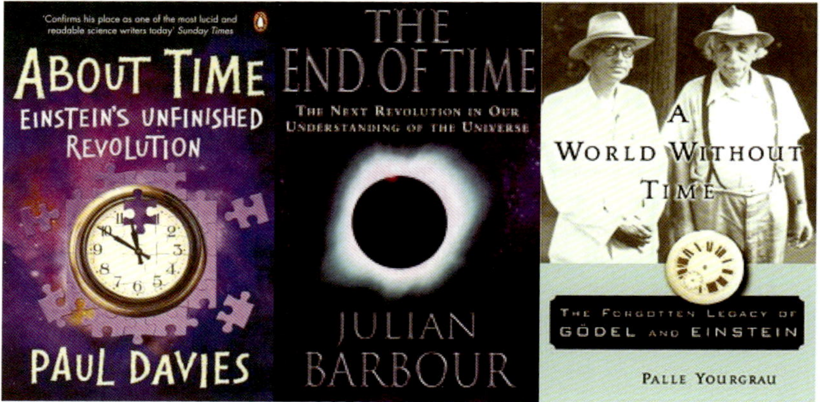

Figure 144 – About Time, The End of Time, A World without Time

I also came across a 1999 paper *Quantum Spacetime: what do we know?* by Carlo Rovelli, see http://arxiv.org/abs/gr-qc/9903045, who said "At the fundamental level we should, simply, forget time". There's also *Process Physics* by Reginald T Cahill, see http://www.scieng.flinders.edu.au/cpes/people/cahill_r/HPS13.pdf dating from 2003. Thank you all. Much of what I learned went into a new improved version, which is pretty much what you've read here if you include *A BRIEF HISTORY OF RELATIVITY*. It's just a rehash of an old idea, but something about it was exciting. I'd stumbled across something that Gödel and Einstein had worked out, I was onto something and my brain was racing. Understanding time seemed to be the key that opened all the doors in physics. If time isn't what you thought, where does that leave spacetime? What does that do to curved spacetime, and what does *that* do to gravity? It was like pulling a loose thread, and all this stuff comes tumbling out. But anyhow, whilst *TIME EXPLAINED* came first, I've pushed it back a few chapters because it's easier that way.

ENERGY EXPLAINED also came out in November 2006. I was getting into the swing of things, thinking about kinetic energy and chemical and nuclear and electromagnetic energy, working backwards from a thrown baseball. I kept seeing elasticity, and springs letting go. And what do springs have? Aha. We are talking stress and tension. I didn't get my ideas about energy from anywhere in particular, and it's no great shakes, because everybody's heard of the matter/energy stress tensor. It's there, as obvious as the nose in front of your face when you sit down and think about it. What surprises me is how obvious it is, and yet when you search the internet for *What is Energy?* all you get is *The Capacity to do Work.* Incredible.

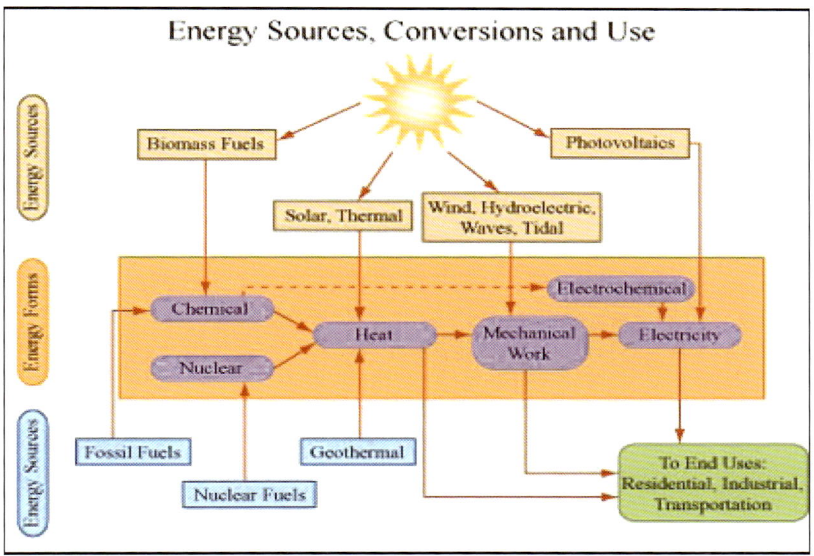

Figure 145 – Energy

The obvious thing to do next was MASS EXPLAINED. People, physicists included, don't seem to understand it. I started doing some research and was pleased to read about "The Mystery of Mass" on the internet. By now I was turning into quite the detective, getting better at interpretation and reading between the lines, all the while getting a little more grasp and getting a feel for things. One guy who was invaluable for my background reading was Peter M Brown, aka pmb. He's a physicist who has written a large number of papers and articles which you can read on his excellent website at http://www.geocities.com/physics_world/. It's chock full of good stuff, such as *Inertial Energy vs Mass, Mass Energy Equivalence, Mass is variable,* and *On the Concept of Mass in Relativity*. Thanks Pete. After reading all Pete's stuff I felt I had an initial grasp of the subject and was ready to take it further. I asked myself *What is Mass?* and slept on it. Then I came across *The Other Meaning of Special Relativity* by Dr Robert A Close, see http://home.att.net/~SolidUniverse/Relativity/Relativity.html. He wrote this paper in 2001, and talked about matter consisting of waves which propagate at the speed of light. I think he deserves a medal. I also found the 1997 paper *Is the electron a photon with toroidal topology?* by J G Williamson and M B van der Mark. John Williamson is at Glasgow University Department of Electronics & Electrical Engineering, and Martin van der Mark is at Philips Research Laboratories in the Netherlands. These guys also deserve a medal, because as soon as I saw the picture of the möbius-doughnut electron I knew intuitively what mass was. After that everything just seemed so obvious. I wrote the essay in December 2006, put it up for feedback, and nobody could bust it. Result.

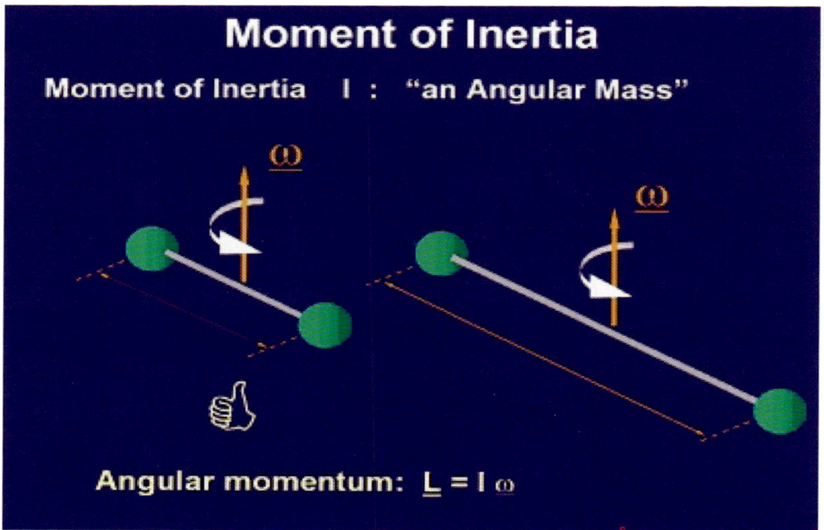

Figure 146 – Angular mass

I wrote *CHARGE EXPLAINED* in April 2007. I thought I'd worked this one out all on my own and sat on it for a couple of months thinking I had a hot property on my hands. Try googling for *What is Charge?* and you'll know what I mean. The only thing that gave me a clue when I was mulling it over, was something on a Science Hobbyist website by William J Beaty, a research engineer at the University of Washington in Seattle. It dates from

1996. Read it yourself at http://amasci.com/elect/charge1.html. What stuck in my mind was that charge is silvery. That was enough, because I had a head start. I understood mass. And once you understand mass as something that you can create and destroy via pair production and annihilation, it's quite easy to take a fresh look at charge. Especially if you've bumped into *The Falaco Soliton: cosmic strings in a swimming pool* by R M Kiehn dating from 2001. See http://arxiv.org/ftp/gr-qc/papers/0101/0101098.pdf. I tried it out in my pond in the back garden, and it wowed me. I felt confident I got it right because electricity, magnetism, refraction, superconduction, and even the electron itself all seem to work out quite logically.

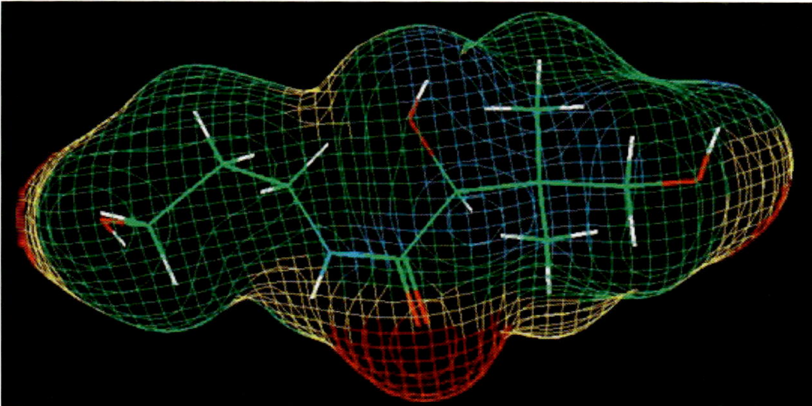

Figure 147 – Molecular electrostatic potential

But whilst I thought I'd worked it out for myself, when I check back, I realise I didn't. I found a forum post from Good Elf dated October 9th 2006, and there it is: *"I think charge is not fundamental. It is partially expressed in this reference: Is the electron a photon with toroidal topology by J.G. Williamson and M.B. van der Mark"*. Special thanks Good Elf. This predates my essay on mass, and I realise that some of my ideas are regurgitated reheats, and I didn't know it. I was rather surprised to realise that *charge is twist* was less novel than I thought, and a little surprised to realise that what I thought was my original thought, wasn't. It's just a synthesis of things I've read, and maybe ignored at the time, and maybe even forgotten. Things like Robert Close's November 2001 paper *Is the Universe a Solid?* on http://home.att.net/~SolidUniverse/. There's Kevin Brown's *Thomas Precession* mathpages article at http://www.mathpages .com/rr/s2-11/2-11.htm. And Robert E Galloway's *Refracting Saddle Wave*

Model of Stable Fundamental Particles at http://www.wbabin.net/science/galloway.pdf. There's also D T Froedge's *The Concept of mass as Interfering Photons,* see http://www.arxdtf.org/. There's Christoph Schiller with *Does Matter Differ from Vacuum?* see http://arxiv.org/PS_cache/gr-qc/pdf/9610/9610066v1.pdf. There's Dave Lush and his *Classical Explanation of Atomic Stability* at http://home.comcast.net/~d.lush/dave_30DecY6.htm. Loads of people. I'm sorry if there's some I've forgotten. Thanks everybody.

REFERENCE FRAMES started life as a discussion on a forum in May 2007 with a guy called Zanket. He's a good bloke but we don't quite see eye to eye. We both think relativity has some issues, but he sees them as fatal flaws whilst I see them as misinterpretations that offer scope for improvement. Anyhow, we were talking about gravity, and I ended up expanding a series of posts into a handy little essay. Peter M Brown's paper *Einstein's Gravitational Field* was very useful for this. See http://xxx.lanl.gov/ftp/physics/papers/0204/0204044.pdf. Thanks again Pete. And thanks to Nick Strobel at http://www.astronomynotes.com. And Rod Knave for his *HyperPhysics* website at Georgia State University on http://hyperphysics.phy-astr.gsu.edu/hbase/hph.html. Thank you all.

I actually wrote *GRAVITY EXPLAINED* after mass but before charge and reference frames. It was R L Collins who got me on to it when I read his December 2000 paper *Mass is Variable* on http://arxiv.org/abs/physics/0012059. It explains so much, particularly if you've also read Robert Close saying *it has recently been shown that torsion waves in an elastic solid are described by a Dirac equation.* It was all falling into place, especially with *The Elastic Continuum Theory of Nature* by G S Sandhu dated September 2002, see http://www.geocities.com/gssandhu_1943/index.html. There's also Reg Norgan's Aether Theory on http://www.aethertheory.co.uk/pdf RFN/Aether_Why.pdf dated 2004. I knew about Einstein's 1920 Leyden address so I wasn't scared of aether. And there tucked away in Reg's aether theory is *why things fall down.* I also found Bruce Harvey's website on http://users.powernet.co.uk/bearsoft/GravP.html. He was telling us about "bracing" within *Gravitational Mass and Attraction as Electromagnetic Phenomena* back in 1997. That's twelve years ago, and I'd never heard of him. I'd never heard about any of these guys. Instead I was getting parallel worlds and time travel and instantaneous action at a distance shoved down my throat. I only found these guys because I went looking. I kept my eyes open and I knew what I was looking for. I'd been talking about energy in terms of stress. I've used a bow and arrow. I know that you don't see stress unless there's tension holding it in place. It's obvious that when you take the time out of spacetime, it's like the derivative of a curve. It's not so much curved spacetime as the gradient in space that opposes localised matter-energy stress. There's a whole pile of people who've been telling us this for years, but all we ever hear about is *spacetime curvature,* even though nobody can explain it to their grandmother. Amazing. But whatever, I can't claim much credit for gravity. These other guys can. I don't know who got there first, but I wrote my *GRAVITY EXPLAINED* essay in February 2007 so I know I didn't. Maybe it was somebody like Dirac, in a paper I've never seen. But anyway, thanks everybody, and thanks to forummers like korosten and Wulf and kjw and many many others.

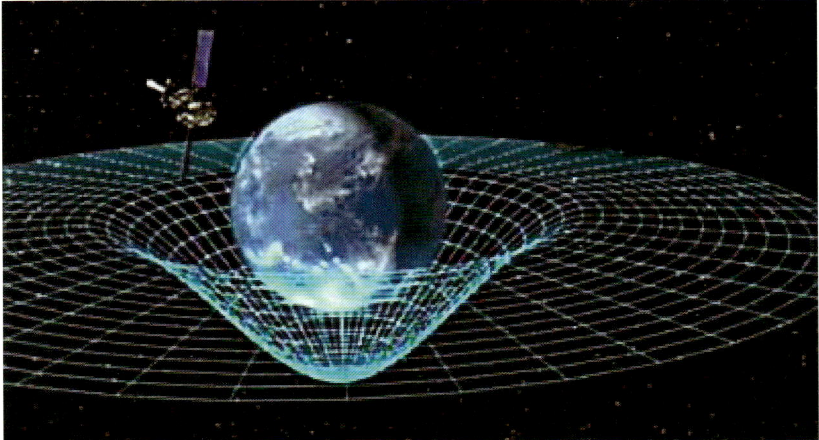

Figure 148 – Gravity mesh

SPACE EXPLAINED began in May 2007. I had a suspicion about it as far back as energy, and a definite clue when I was doing gravity. The key phrase is *the world is painted in light.* I had my head full of twist and was tussling with polarization as in poles of a battery as well as the polarization of light. But I found myself getting pulled in two directions, namely particle physics and cosmology. Initially I started writing about space, black holes, and the universe in a single essay, but it started getting too big. So I decided that *SPACE* should be more about the stuff of space and energy rather than cosmology. So I hacked the space out into a separate piece which was ready by June 2007. I'd like to thank Steven Dutch, a geologist at the University of Wisconsin for his website <u>http://www.uwgb.edu/DutchS/petrolgy/gen light.htm</u> which talks about polarized light. In addition I'm grateful to Rod Nave again for the *Polarization Concepts* pages on <u>http://hyperphysics</u>

.phy-astr.gsu.edu/hbase/phyopt/polarcon.html. And thanks Chris Hillman, Turbo, HallsOfIvy, Pervect, and Marcus for the answers and pointers and leads.

Figure 149 – Silly putty in a vacuum chamber

PARTICLES EXPLAINED dates from August 2007. I was conscious I had something that I could take further, and had been challenged to do so on an internet forum. I'd already touched on particles in previous essays, and I'd previously written bits and pieces about things like *the crease in my pants* along with whirlpools and waves and the nature of matter. So I was up for it. I like to think I joined the dots on this one and came up with the notion that particles are just knots of distance variation. But I wouldn't put money on it. I'm sure I've read the phrase *extended entities* somewhere before, but I'm sorry, I can't remember where. I was encouraged by Vernon Brown, who I bumped into in May 2007, and his *Photon Theory*. You can read it online at http://photontheory.com/pte.html. When I searched the internet, up came http://libinfo.uark.edu/specialcollections/findingaids/pryor/g2s07.asp telling me Vern did this in 1992. Seventeen years ago! Thanks Vern, and sympathies. You're a kindred soul, and you gave me encouragement, plus a lead to Robert Kemp which gave me a big clue about Planck's constant and the ultimate quantum. He was talking about a constant electromagnetic amplitude in his 1994 paper *The Quantization of Electromagnetic Change* dating from 1994 on http://photontheory.com/Kemp/Kemp.html. That's fifteen years ago. OK, I sussed the constant distance, but I wouldn't be surprised if he did that years back. Thanks Robert and Vern. I ended up beefing up the punch line and putting in a little history to turn it into its own chapter called *QUANTUM MECHANICS*.

MORE PARTICLES was more of the same. The challenge was all to do with the standard model, and initially that's what I called it. I ended up splitting out *THE STANDARD MODEL* for readability. I looked at a whole pile of websites, including Rod Nave's HyperPhysics again and I worked out the trefoil knot and the neutrinos by my own sweet self. But I wouldn't be surprised if somebody rattled off half a dozen guys who got there first, or dug out a fully-developed theory that's been gathering dust for twenty years. I dug up *Energy is Everything, a Quantum Explanation of Gravity and Inertia* by David W Talmage and Richard J Sanderson dated 2002. See http://www.geocities.com/davidwtalmage/quantum2.htm. I also found *The Origin of Gravity* by Albrecht Giese dated June 2004. See http://ag-physics.org/gravity/. It's that same gravity explanation again, complete with the variable c and refracting downward motion. And here we go, up pops *A Look at the Abandoned Contributions to Cosmology of Dirac, Sciama and Dicke* by Alexander Unzicker. See http://arxiv.org/abs/0708.3518, where the abstract says *We study Dirac's article on the large number hypothesis (Proc.Roy.Soc. Lon. A 165, 199; 1938), Sciama's proposal of realizing Mach's principle (MNRAS 113, 34; 1953), and Robert Dicke's considerations on a flat-space representation of general relativity with a variable speed of light (Rev.Mod.Phys. 29, 363; 1957).* Why didn't I know about Dicke's variable speed of light? And why did I only hear about Puthoff and http://en.wikipedia.org/wiki/Polarizable_vacuum from Norman Albers? See his *Gravitation and Vacuum Polarization* on http://laps.noaa .gov/albers/physics /na/. I don't know. But what I do know is that you usually don't hear about this material unless you know what you're looking for. Then you go looking for it, and it's everywhere you look. Like Sundance Bilson-Thompson with *You Are Made of SpaceTime*. You spot an image that looks familiar and wham, up comes Lew Price and his *Nether Theory*. See http://www.promedia.net/users/greebo/price.htm. You read Lee Smolin's *Trouble With Physics* and there on page 105 paragraph one he says mass is the measure of a particle's energy when it's motionless. I've just found a website http://classicalmatter.com/ClassicalMatterWaves.html. It's Robert Close again, writing a book called *A Classical Theory of Matter Waves*, and I'm wondering why he's not a household name. And who's Harold Aspden? He lives near me. What's he been saying and for how long on http://www.aspden.org/, and why don't I know about him? You just don't spot this sort of thing until you already know what you're looking for. So I expect there's half a dozen guys who've been saying particles are knots for the last ten years, and the rest. But anyhow, *MORE PARTICLES* wasn't an internet essay, but instead it was a section of a formal scientific paper *A Qualitative 3+1 Dimensional Geometrical Model.* I finished it in October 2007 and put up on the internet for feedback and review prior to submission. In essence the paper was a condensed version of the essays to date, plus extra sections that ended up being chapters of this book.

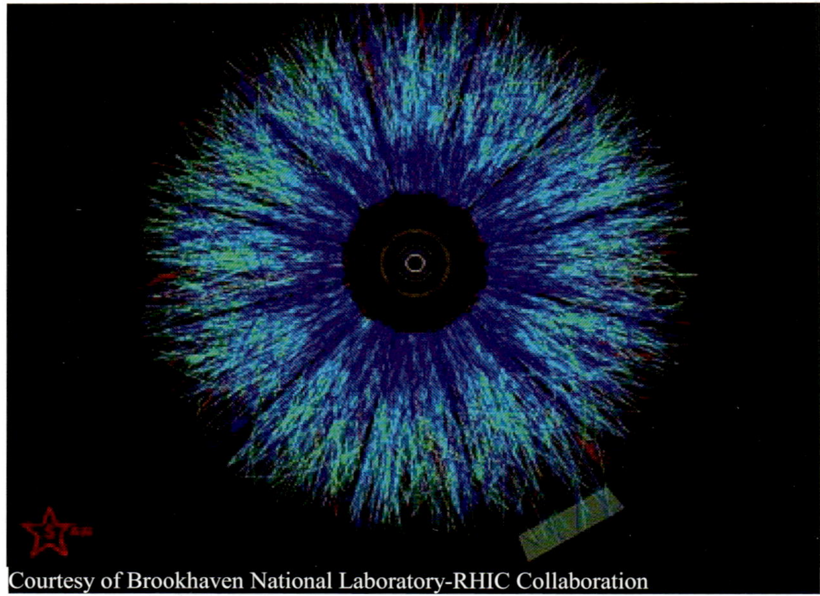

Figure 150 – STAR particle collision

THE UNIVERSE started life back in May 2007, but didn't emerge in its present form as a freestanding chapter until early 2008. The salient points seemed to flow naturally from everything else so far. You see how lambda is an inherent property of compressed elastic space. But since it's tied up as matter in some locations, space just can not be homogeneous, and that's bound to threaten dark matter with a MOND-like gravity that doesn't quite follow the inverse square rule. And rather than a hypersphere, we can conceive of a hyperbolic-type universe beyond which there is no more distance. It all seems fairly obvious, trivial even, though of course it really needs some rigor to put some flesh on the bones. But still, I couldn't have done it without all the excellent material provided by others. There are some absolutely brilliant websites out there, including http://onserve.arc.nasa.gov, http://universe-review.ca, http://www.astronomyonline.org, http://www.ast-ronomy.ohio-state.edu/, http://chandra.harvard.edu, http://www.physics.ucla.edu/demoweb/, http://hubblesite.org/, and many many others. Thank you everybody involved.

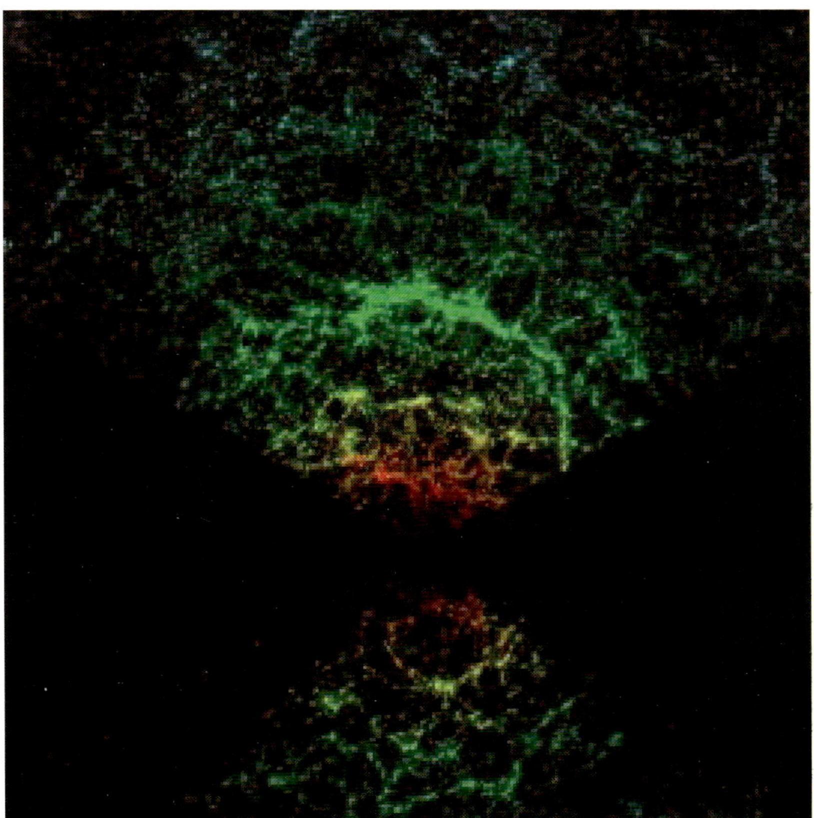

Figure 151 – The universe (SDSS)

BLACK HOLES and *THE BIG BANG* were similar. I owe thanks to the same set of people, and more. It was May 2007, I was talking to Zanket and Andrew Grey about singularities and frozen stars and what Schwarzschild really meant. I found myself re-reading *About Time* by Paul Davies again.

The infinite time dilation jumped out, and before I knew it I was tussling with something and nothing and had a whole new topic on my hands. I learned a great deal from excellent internet sources such as NASA's http://www.hubblesite.org, Rick Murphy's http://www.astronomyonline.org, Matt McIrvin's Black Hole FAQs at http://antwrp.gsfc.nasa.gov/htmltest /gifcity/bh_pub_faq.html, and Adam D Helfer's *Do Black Holes Radiate?* on http://xxx.lanl.gov/PS_cache/gr-qc/pdf/0304/0304042v1.pdf. Can I also say thanks to Kevin Brown for *The Formation and Growth of Black Holes* on http://www.mathpages.com/rr/s7-02/7-02.htm. The part saying *the two most common conceptual models for general relativity have been the "geometric interpretation" (as exemplified by Misner/Thorne/Wheeler's "Gravitation") and the "field interpretation" (as in Weinberg's "Gravitation and Cosmology")* was news to me.

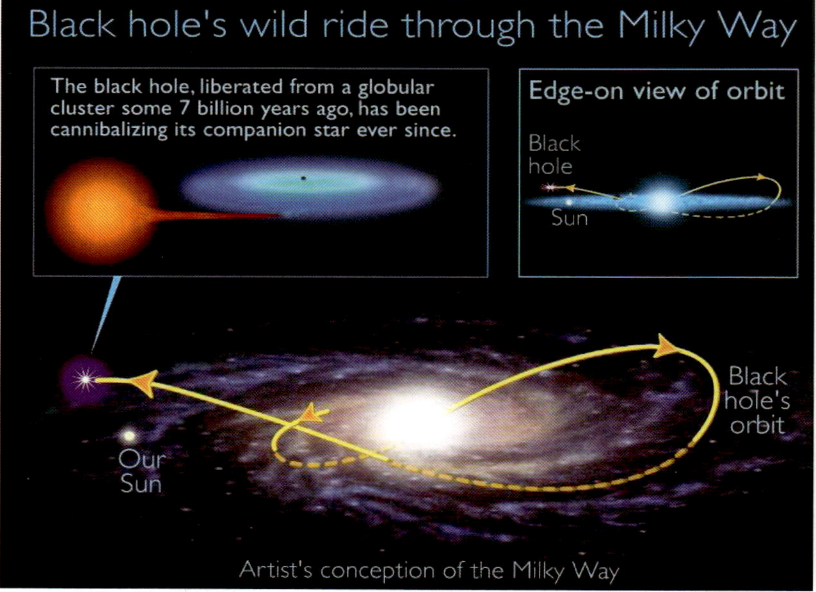

Figure 152 – Black-hole microquasar XTE J1118+480

That's about it. There's other bits and pieces written at various times, some of which grew out of other chapters. *MONEY* started life as a paragraph in energy. *GOING NOWHERE FAST* came out of mass, whilst *A NOTE ON NEWTON* and *EINSTEIN'S HEROES* came out of gravity. *THE FINE STRUCTURE CONSTANT* along with *INFLATION COMES FOR FREE* is something recent, written while I was pulling the rest together, whilst the *RELATIVITY+* chapter is just a round up. And to round up, I also owe thanks to all the historical figures I've mentioned in this book. There's Einstein, Feynman, Dirac, Planck, De Broglie, Schrodinger, Lorentz, Poincaré, Voigt, FitzGerald, Weyl, Gödel, Tesla, Millikan, Maxwell, Faraday, Newton, and more besides. They've all passed into history, so I can't thank them directly. But I can thank them publicly: thank you all. And Aristotle, and Galileo, and Copernicus and Bruno and all the rest. Thanks for being curious, and dedicated, and determined, and lighting the way for those who follow. And thank you to those who do follow, scientists and science writers everywhere. Thank you wiki and website contributors, thank you everybody I've mentioned in my references and figures, thank you Esther Allerton for the proofreading, thanks everybody I've talked to on websites and forums and email. And thanks to all the people who gave permission to use the images, and to those who created them in the first place.

The one thing I've learned from all this is that a lot of people have contributed a lot of work, and sometimes they get no recognition for it. Yes, sometimes they achieve fame, but sometimes the world of science just passes them by. People ignore what they say because it doesn't fit the paradigm. They dismiss it, and that's why the mystery persists. This has been going on for a "long long" time. Too long. The answer is there, but you just don't see it until you learn to recognise it. And then, quite amazingly, it's everywhere you look. Now I keep stumbling across what looks like an elephant in the room that nobody told me about. Now I've seen the elephant, I recognise it as *the same elephant*. It's the same simple idea of the vacuum modelled as elastic. I never heard about it until I started doubting things like String Theory and Boltzmann Brains and working it all out. I think it's because aether is some kind of dirty word, almost a taboo, and there's some kind of running battle at the heart of science. I think it's something to do with the nature of consciousness, something psychological, maybe sociological, and I can only describe it as the schism between thinking and learning. As scientists we hone our logic by using our logic. We learn how to think by *thinking*, not by *learning*. And by thinking, we develop a new mental model and discover a new truth. But this new truth must challenge the old truth, the learned truth, and so we enter into conflict. It can be fierce. Reputations are at stake, lifetimes of work. It's fierce enough to put people beyond the pale, and elbow Einstein and Dirac out of the mainstream. It's a battle, and in that battle the new truth scurries in the shadows like early mammals, kept in the dark by groupthink and conviction and vested interest. By people with funding to preserve, supported by a peer review system that lets experts in the field stifle the competition.

Most scientists regarded the new streamlined
peer-review process as 'quite an improvement.'

Figure 153 – Peer review process

But I like to think that in the end, thinking wins. The truth will out because once you discover a truth, you are better able to see it. You see it in all its various guises, dressed in different language and different terminology. You see the common ground and home in on it rather than fighting over the bones of contention. You know that despite the different approaches and the flaws, the beautiful simple truth is in there shining through. Once you were blind to it, but now it's oh so obvious. You can spot it a mile off. And you wonder why you couldn't see it before.

I see the world of physics as something like a cocktail party. Many people can't see the elephant in the room. They don't notice that some of their colleagues are clustered around this elephant, grasping it, getting a feel for it, thinking about it. Some are running their hands experimentally over the trunk, puzzling at what they've found. Others are touching the ears, whilst others probe the tail. They come up with different theories concerning this elephant, because they can't see it properly, because they're wearing dark dark glasses, and they can't take them off. They get parts of the description right. They get other parts wrong, and that's enough for rejection so they don't get an airing. But once you've seen the elephant, you understand what people are talking about. You understand that it's *the same elephant*. You can spot it a mile off. And you can also spot that there's more people around the elephant every day.

They're forming a crowd, and the crowd is growing. People are starting to piece it together, because the internet is helping so much. It's been tough going, because people are human, they have their pride and their conviction and their funding at stake. And String Theory has had physics tied down like Gulliver in Lilliput. But Lilliput is where I come in. It's a real place, and that's where I'm from. They can't tie *me* down. I can cut loose.

Figure 154 – Gulliver

I'm not a religious sort of a guy, but if I were holding a knife ready to cut the rope and unleash the sleeping giant, I know exactly what I'd say. It's kind of a *Sunshine* kind of a moment. You know, right at the end. The payload has started to scintillate but the sun has crashed in. And then there's that curious pause, and Capa stretches out his hand:

Let there be light.

REFERENCES

[1] *Classical unified field theory*
See the American Institute of Physics (AIP) website http://www.aip.org/ history/einstein/index.html which gives an excellent Einstein history. You can view parts of papers such as *A New Form of the General Relativistic Field Equations* written in 1955 with assistant Bruria Kaufman. Also see the Wikipedia article on classical unified field theories at http://en.wikipedia.org /wiki/Classical_unified_field_ theories.

[2] *Tom Kilburn, godfather of Computer Science*
See http://www.computer50.org/mark1/kilburn.html maintained by Doctor Brian Napper of the Department of Computer Science at the University of Manchester.

[3] *Quantum Mechanics*
For beginners I recommend *Introducing Quantum Theory* by J P McEvoy illustrated by Oscar Zarate. It was originally published as *Quantum Theory for Beginners* in 1996 by Icon books.

[4] *The Pleasure of Finding Things Out*
This is the title of a Richard Feynman book edited by Jeffrey Robins, foreword by Freeman Dyson, published by Perseus Publishing in 1999.

[5] *He went to Caltech instead*
It was because of the weather, and because he wanted to stay connected and teach, which Einstein thought was a drag. See http://en.wikipedia.org/wiki/ Richard_Feynman and *Surely You're Joking Mr Feynman!* published by Norton in 1985, which came out of interviews with Ralph Leighton. It was edited by Edward Hutchings.

[6] *Robert Millikan*
See biographical details at http://nobelprize.org/nobel_prizes/physics/laureates /1923/millikan-bio.html. Also see *Einstein Redux* by Jane Dietriech at http:// eands.caltech.edu/articles/EinsteinFeature.pdf re Judith Goodstein's 1991 book *Millikan's School* and the Einstein papers project.

[7] *O-rings*
Feynman was on the Rogers Commission investigating the Challenger Accident, and famously demonstrated that O-rings lose elasticity using a glass of iced water. Read what he said in *Volume 2: Appendix F - Personal Observations on Reliability of Shuttle by R. P. Feynman* at http://history.nasa .gov/rogersrep/v2appf.htm. Also see the Wikipedia article http://en.wikipedia. org/wiki/Space_Shuttle_Challenger_disaster along with *What Do You Care*

What People Think? published by Norton in 1988, which also came out of interviews with Ralph Leighton.

[8] *On the ElectroDynamics of Moving Bodies*
This is Einstein's famous 1905 paper. You can find an online translation at the fourmilab website maintained by John Walker. See http://fourmilab.ch/etexts /einstein/specrel/www/.

[9] *The Mystery of the Einstein–Poincaré Connection*
See Olivier Darrigol's essay at http://www.journals.uchicago.edu/doi/full/10. 1086/430652?cookieSet=1. If you can read French you'll find stuff which is far more hard hitting. Also see http://en.wikipedia.org/wiki/Relativity_priority_ dispute.

[10] *Presentism*
This is a philosophical concept introduced by John McTaggart in 1908 in an essay called *The Unreality of Time,* published in the journal MIND. You can read it at http://www.ditext.com/mctaggart/time.html. The opposing concept is called Eternalism, and involves the block universe and problems for free will.

[11] *Space and Time*
If you can read German, you can read Minkowski's original paper at http: //de.wikisource.org/wiki/Raum_und_Zeit_(Minkowski). I can't find an English translation online, but it starts on page 73 of *The Principle of Relativity: A Collection of Original Memoirs on the Special and General Theory of Relativity* published by Dover in 1952.

[12] *Since the mathematicians pounced on the relativity theory...*
This quote is in Albrecht Folsing's 1998 Penguin book *Albert Einstein: A Biography* translated by Ewald Osers, see page 261. Search the internet and you'll come up with *David Hilbert and the Axiomatization of Physics,* which says on page 224 that *such a statement appears in Seelig 1954.* Ths is Ideas and Opinions By Albert Einstein edited by Carl Seelig, translated by Sonja Bargmann, and published by Bonanza books in 1954.

[13] *The Foundation of The General Theory of Relativity*
You can find this in *The collected papers of Albert Einstein, volume 6, the Berlin years, writings 1914-1917, document 30.* It's translated by Alfred Engel, and edited by A. J. Kox, Martin J. Klein, and Robert Schulmann. It was published by Princeton University Press in 1997, and you can read it online on http://www.alberteinstein.info/gallery/pdf/CP6Doc30_English_pp146-200.pdf.

[14] *David Hilbert, who had been close to Minkowski*
David Hilbert was close to Minkowski, and like Minkowski, thought mathematics was of paramount importance. I pause for thought when I read Otto Blumenthal saying *when it comes to penetrating insight, only a few of the*

very greatest were the equal of Hilbert. There was quite a tussle going on at the time between what was known as theoretical physics and mathematical physics. Einstein was in the former camp, Minkowski and Hilbert in the latter. Einstein became more mathematically inclined as he got older, but arguably at the cost of the insight that served him so well in his early years. And yet Weyl was Hilbert's student. I can only wonder at what might have been.

[15] *Relativity: The Special and General Theory*
This short book by Albert Einstein was originally written in 1916 and published by Vieweg in Germany, then translated by Robert Lawson and published in English by Methuen in 1920. You can buy it from various sources or download it for free from Project Gutenberg, see http://www.gutenberg.org/etext/5001.

[16] *Leyden address*
The title of this is *Ether and the theory of relativity.* It's online at various web sites, such as http://www.zionism-israel.com/Albert_Einstein/Albert_Einstein_Ether_Relativity.htm. Lorentz is mentioned quite a lot, an indication of Einstein's good opinion of him.

[17] *Gödel's Incompleteness Theorem*
Some folk use it to say things like *we can never create artificial intelligence,* but that's not what it's about. It's to do with axioms and proof, saying you can't always trust the things you take for granted. See http://en.wikipedia.org/wiki/G%C3%B6del's_incompleteness_theorems for a good article.

[18] *time, as we ordinarily understand it, does not exist.*
This paragraph is from the front flap of Palle Yourgrau's book *A World Without Time: The Forgotten Legacy of Gödel and Einstein* published by Basic Books in 2005. Palle Yourgrau is professor of philosophy at Brandeis University in Boston.

[19] *perhaps the concept of field was inadequate*
See *Unified Field Theory and Einstein* by S C Tiwari dating from 2006. You can view it online at arxiv, see http://arxiv.org/ftp/physics/papers/0602/0602 112.pdf. Note the quote: *Einstein, in his last paper on the subject, admitted that perhaps the concept of field was inadequate for the unified theory which he was seeking.*

[20] *You must not fool yourself, and you are the easiest person to fool*
See http://en.wikiquote.org/wiki/Richard_Feynman. It says it's from a lecture *What is and What Should be the Role of Scientific Culture in Modern Society?* given at the Galileo Symposium in Italy 1964.

[21] *Psychology of Belief*
I'm not sure where this started, but the journal MIND featured a paper *The Psychology of Belief* by Professor William James in 1889. It's on JSTOR which

is not freely available, but see http://www.jstor.org/pss/2247316. Also have a browse through Christopher D Green's *Classics in the History of Psychology* at http://psychclassics.yorku.ca/author.htm.

[22] *Thomas Kuhn*

He wrote *The Structure of Scientific Revolutions* published by Chicago University Press in 1962. See the synopsis by Frank Pajares on http://www.des. emory.edu/mfp/kuhnsyn.html which starts: *A scientific community cannot practice its trade without some set of received beliefs. These beliefs form the foundation of the "educational initiation that prepares and licenses the student for professional practice". The nature of the "rigorous and rigid" preparation helps ensure that the received beliefs are firmly fixed in the student's mind.*

[23] *Donald Rumsfeld was right*

I'm talking about the famous "don't know" quote from the Department of Defence news briefing of 12 February 2002: *Reports that say that something hasn't happened are always interesting to me, because as we know, there are known knowns; there are things we know we know. We also know there are known unknowns; that is to say we know there are some things we do not know. But there are also unknown unknowns - the ones we don't know we don't know.* I know what he means.

[24] *The release of ADP and inorganic phosphate...*

This quote is from the English Wikipedia article *Muscle Contraction* at http://en.wikipedia.org/wiki/Muscle_contraction. Also see *Molecular Biology of the Cell* by Alberts et al, published by Garland, page http://www.ncbi.nlm.nih.gov/books/bv.fcgi?highlight=myosin&rid=mboc4.section.3041#3065, to read more about molecular motors.

[25] *Cubane*

This quote is from a paper supplied by Professor Emeritus Philip Eaton of the Department of Chemistry at Chicago University. The formal reference is *Eaton, P. E.; Zhang, M.; Gilardi, R.; Gelber, N.; Iyer, S.; Surapaneni, R.; "Octanitrocubane: A New Nitrocarbon", Propellants, Explosives, Pyrotechnics, 27, 1-6 (2002).*

[26] *Vorton*

See *Stationary ring solitons in field theory – knots and vortons by Eugen Radu and Mikhail S. Volkov* at http://arxiv.org/PS_cache/arxiv/pdf/0804/0804.1357 v2.pdf. They're from the Laboratory of Mathematics and Theoretical Physics at the University of Tours.

[27] *Toroidal Soliton*

I'm referring here to a paper by John G Williamson and Martin B van der Mark entitled *Is the electron a photon with toroidal topology?* It appeared in the journal Annales de la Fondation Louis de Broglie, within volume 22 in 1997.

You can view it online at http://www.cybsoc.org/electron.pdf along with a preprint of John's new paper *On the nature of the electron and other particles* at http://www.cybsoc.org/electremdense2008v3.pdf.

[28] *Falaco Soliton*
You really must try out "playing with plates". See *Kiehn, R. M.; (2004 - 2nd edition 2007) Cosmology, Falaco Solitons and the Arrow of Time, "Non-Equilibrium Systems and Irreversible Processes Vol 2", Lulu Enterprises Inc., 3131 RDU Center, Suite 210, Morrisville, NC 27560. URL: http://www.lulu. com/kiehn.*

[29] *Kinetic Theory of Gases*
There's various places where you can look at this, such as Michael Fowler's *Kinetic Theory of Gases: A Brief Review*. He's a physics professor at the University of Virginia, see http://galileo.phys.virginia.edu/classes/252/kinetic _theory.html. Note Bernoulli talking about elastic.

[30] *Luca Turin*
I came across Luca Turin in my *New Scientist* magazine, which I tend to read on Saturday mornings at the breakfast bar with a cup of coffee. Here's the teaser for the article from the 18th November 2006 issue: *Dubbed "the Emperor of Scent", biophysicist Luca Turin has stirred up a heated debate about the inner workings of our sense of smell. In the mid-1990s he revived the argument that the human nose detects the presence of a compound from its molecular vibrations rather than its shape.*

[31] *GPS*
To find out the real gen on GPS and relativity see: *Ashby, N.; "Relativity in the Global Positioning System", Living Reviews in Relativity, Volume 6, 2003-1. URL: http://www.livingreviews.org/lrr-2003-1.* Note the section where he says: *In order for the satellite clock to appear to an observer on the geoid to beat at the chosen frequency of 10.23 MHz, the satellite clocks are adjusted lower in frequency so that the proper frequency is [1 - 4.4647 x 10^{-10}] x 10.23 MHz = 10.229 999 995 43 MHz.*

[32] *Pound-Rebka Experiment*
The original paper is *Pound, R. V.; Rebka Jr. G. A.; "Gravitational Red-Shift in Nuclear Resonance", Physical Review Letters 3: 439-441 (1959).* Professor Robert Pound and Doctor Glen Rebka got a medal for this, and deservedly so. I'm not so keen on their phrase *the weight of light* myself, but it does touch on the deep and elegant relationship between momentum and inertia.

[33] *Shapiro Effect*
This is also know as the Shapiro Time Delay, and was discovered in 1964 by Irwin Shapiro at MIT before he became a professor. Bounce a radar beam off Mars, and it takes a little longer to come back if it's got to skirt the Sun. People

say this is because of "curved spacetime", but it's really because the gravitational time dilation means it goes slower near the Sun. The original paper is *Shapiro, Irwin, I.; "Fourth Test of General Relativity", Physical Review Letters 13: 789-791 (1964)*. I can't find it online I'm afraid. But you can see various of his other papers on arXiv: http://arxiv.org/find/astro-ph /1/au:+Shapiro_I/0/1/0/all /0/1.

[34] *Variable speed of light*
John Moffat was the first physicist in modern times to talk about a variable speed of light. That was in 1992, but he was largely ignored. He's Emeritus Professor of Physics at Toronto, and thankfully people now pay more attention. See the Wikipedia article on him at http://en.wikipedia.org/wiki/John_Moffat_ %28physicist%29 and his paper *Moffat, J.; "Superluminary Universe: A Possible Solution to the Initial Value Problem in Cosmology", Int.J.Mod.Phys. D2 351-366 (1993). URL: arXiv:gr-qc/9211020v2*. A few years later in 1998 Joao Magueijo and Andreas Albrecht had similar ideas, not knowing about Moffat. See *Albrecht, A.; Magueijo, J.; "A time varying speed of light as a solution to cosmological puzzles", Phys.Rev. D59 (1999) 043516. URL: arXiv:astro-ph/9811018v2*. Here's what Wikipedia says: *"Their paper made it into the more prestigious journal, Physical Review D, which had rejected Moffat's paper years earlier. When Moffat saw this, he was upset and contacted Magueijo. But after Magueijo realized what had happened, he was quick to give Moffat due credit for having first proposed the idea. In fact, Moffat and Magueijo became friends, and Magueijo even devoted a whole chapter to Moffat in his 2002 book titled "Faster Than the Speed of Light". After that, the number of physicists citing Moffat's work in academic journals skyrocketed"*. Joao Magueijo is professor of physics at UCL. Check out his book *Faster Than the Speed of Light: The Story of a Scientific Speculation* first published by Heinemann. It's quite critical of modern science, but it tells you what it's like, and is a really good read. He's talking here about c being greater in the early universe, not lower, and he doesn't make the connection between variable c and gravity. But I imagine he does now. See *Magueijo, J.; "New varying speed of light theories", Rept. Prog. Phys. 66 (2003). URL: arXiv:astro-ph/0305457v3*. It's an interesting paper, see this quote: *Ironically, the first "varying-constant" was the speed of light, as suggested by Kelvin and Tait in 1874. Some 30 years before Einstein's proposal of special relativity, a varying c did not shock anyone.*

[35] *Einstein's Gravitational Field*
Peter M Brown wrote a somewhat historical physics paper of this name highlighting how the modern interpretation of general relativity is not in accord with Einstein's view. See *Brown, P. M.; "Einstein's gravitational field", URL: arXiv:physics/0204044v2*. This is from the abstract: *The interpretation of gravity as a curvature in space-time is an interpretation Einstein did not agree with*. What comes across strongly is that a uniform gravitational field is a contradiction in terms that people just can't see. Pete quotes John R Ray saying:

The first thing to note about the 1911 version of the principle of equivalence is that what in 1911 is called a uniform gravitational field ends up in general relativity not to be a gravitational field at all - the Riemann tensor is here identically zero. Also check out Pete's website at http://www.geocities.com/physics_world, which is full of interesting material.

[36] *If only you know how to look*
See Clifford M Will's paper *"The Confrontation between General Relativity and Experiment"*, Living Rev. Relativity 9, (2006), 3. URL: *arXiv:gr-qc/0510072v2*. The experiments to test general relativity are observations. And look out for this on page 6: *the steady accumulation of experimental support, together with the successful merger of special relativity with quantum mechanics, led to its being accepted by mainstream physicists by the late 1920s.* Clifford M Will is Professor of Physics at The McDonnell Center for the Space Sciences, a facility at Washington University St Louis.

[37] *That gravity should be innate, inherent, and essential to matter..*
This famous Newton quote comes from a letter written by him to Dr Richard Bentley on 25 February 1692. You can find it in various online locations, and in *Newton's Philosophy Of Nature: Selections From His Writings* edited and arranged by Horace Standish Thayer and published by Hafner in 1953.

[38] *Dirac Large Numbers Hypothesis*
Dirac goes way back, and I don't do him justice. See http://en.wikipedia.org/wiki/Dirac_large_numbers_hypothesis where you can read: *There is another ratio with this order of magnitude: the ratio of the electrical to the gravitational forces between a proton and an electron, $4\pi\varepsilon_0 Gm_{pr}m_e/e^2$ ($\approx 4.4 \times 10^{-40}$). Hence, taking the charge e of the electron, the mass m_{pr}/m_e of the proton/electron, and the permittivity factor $4\pi\varepsilon_0$ as units, the gravitational constant equals $G = 10^{-40}$.* Dirac interpreted this to mean that G varies with time as $G \propto 1/t$, and built a cosmology out of this idea. Dirac's approach was geometric not algebraic, and he had an aether. He was one of the greats, the founder of quantum electrodynamics, and Lucasian Professor of Mathematics at Cambridge, the position once held by Isaac Newton and currently held by Stephen Hawking. See http://www.lucasianchair.org/20/dirac.html written by Robert Bruen for details such as: *In 1927 Dirac united quantum mechanics and relativity by deriving a relativistically invariant form of Schrodinger's wave equation, predicting positively charged electrons, which were seen later in 1932. He also predicted that antimatter will annihilate when in contact with matter. He is best recognized for pulling together all of the disparate theories of quantum mechanics... He was able to justify the spin of the electron on a theoretical level which had eluded physicists up to this point. It was this insight that separated him from others of his time and won him the Nobel Prize in 1933... The Dirac equation rejected the idea that an electron was a point in space that orbits around the nucleus of the atom.* Also see the *Positron Anihilation Net* website, and in particular http://www.positronannihilation.net

/history.html where you can read: *The positron as the antiparticle of the electron was predicted by Dirac ("On the quantum theory of electron":1928). First experimental indications of an unknown particle were found in cloud-chamber photographs of cosmic rays (Anderson 1932). This particle was identified later as the positron, which was thus the first antiparticle in physics. The annihilation of the positron with electrons in matter was first studied in the 1940s.* The picture below is from 1933 and is his official Nobel Prize portrait. See http://nobelprize.org/nobel_prizes/physics/laureates/1933/dirac-bio.html.

[39] *Are not gross bodies and light convertible into one another?*
This refers to queries 30, 20, and 21, which are on pages 374 and 350 of my copy of *Opticks*, a Cosimo paperback published in 2007. When you understand his thinking, you can understand why Newton was so interested in light. You can also understand why he was interested in alchemy. If you can convert light into matter, turning lead into gold must be a piece of cake. Besides, alchemy was more like chemistry in those days. All the top natural philosophers were doing it. See http://www.pbs.org/wgbh/nova/newton/alch-newman.html where Bill Newman talks sympathetically. Newton was light years ahead of his contemporaries. He was immensely rational and experimental, and is generally thought to be against making hypotheses on account of the quote *I feign no hypothesis*. But you have to read this literally. He didn't *feign* them, pretending that hypotheses were facts, but he did propose them, as evidenced by the queries at the back of Opticks. Also see query 18 where he says *Is not the heat of the warm room conveyed through the vacuum by the vibrations of a much subtler medium than air?* He is generally thought to have rejected the notion of light as a wave, but what he really rejected was the idea of a viscous fluid ether.

[40] *Einstein's last interview*
Einstein's last interview, with Isaac Bernard Cohen, was featured in Scientific American in 1955. See Professor Rob Iliffe's *Newton Papers Project* website at

http://www.newtonproject.sussex.ac.uk/texts/viewtext.php?id=OTHE00009&mode=normalized for more.

[41] *Einstein, his life and universe*
Walter Isaacson's book *Einstein, his life and universe* was published by Simon & Schuster in 2007. Walter Isaacson is the former chairman and CEO of CNN and is now president and CEO of the Aspen Institute.

[42] *Einstein's heroes*
See the book *Einstein's Heroes: Imagining the world through the language of mathematics* by Dr Robyn Arianrhod, initially published by the University of Queensland Press in 2003, then by the Oxford University Press (USA) in 2005.

[43] *strongly elastic, and in a word much like air in all respects*
This is from a letter by Newton to Robert Boyle dated 28 February 1678. I found it on page 1 of *Isaac Newton: Philosophical Writings* edited by Andrew Janiak, printed by Cambridge University Press in 2004. Andrew is a professor of philosophy at Duke University North Carolina.

[44] *Faraday Lectures*
Michael Faraday started the tradition of Christmas lectures. They're on TV these days, and I think they're wonderful. The idea was to educate and excite and it works. Faraday's lectures were real hands-on stuff, chock full of graphic demonstrations. I really wish I could have been present. For me he's right up there with Newton and Einstein. I would encourage you to find out more about Faraday, and I urge you to take the time to read the transcripts of his lectures such as http://www.fordham.edu/halsall/mod/1859Faraday-forces.html.

[45] *Einstein on Faraday*
This is from Einstein's paper *The Fundaments of Theoretical Physics* printed in Science magazine in 1940. You can see it online inside a 2002 paper by Pharis Williams entitled *Energy and Entropy as the Fundaments of Theoretical Physics*. It's at the Molecular Diversity Preservation International website in the entropy section. See http://www.mdpi.org/entropy/papers/e4040128.pdf.

[46] *Maxwell's Dynamical Theory of the Electromagnetic Field*
Maxwell was another of the greats, see http://en.wikipedia.org/wiki/A_Dynamical_Theory_of_the_Electromagnetic_Field along with http://en.wikipedia.org/wiki/James_Clerk_Maxwell. This is interesting: *Maxwell believed that the propagation of light required a medium for the waves, dubbed the luminiferous aether. Over time, the existence of such a medium, permeating all space and yet apparently undetectable by mechanical means, proved more and more difficult to reconcile with experiments such as the Michelson-Morley experiment. Moreover, it seemed to require an absolute frame of reference in which the equations were valid, with the distasteful result that the equations changed form for a moving observer. These difficulties inspired Albert Einstein to*

formulate the theory of special relativity, and in the process Einstein dispensed with the requirement of a luminiferous aether. That's what's generally accepted, but people don't know about Einstein's Leyden address.

[47] *Weyl Transformation*
See William Straub's excellent website http://www.weylmann .com/ and follow the links to http://www.weylmann.com/weyltheory.pdf, http://www.weylmann .com/scalefactor.pdf and http://www.weylmann.com/weyldirac.pdf. I smile wryly when I see *neutrinos exist and are all described by the Weyl spinor* ψ_L *which violates parity.* Look at the ψ symbol. Now think "neutrino bullwhip".

[48] *Einstein's greatest blunder*
It's January 14th 1931 and Einstein is visiting the Carnegie Institution's Mount Wilson Observatory headquarters in Pasadena. Albert isn't smiling this day on account of his greatest blunder. He should have predicted the expansion of the universe long before Hubble spotted it. But instead he believed we lived in a *static* universe, and fudged it with a cosmological constant. See http://www. ciw.edu/ to find out more about Carnegie.

Milton Humason, Edwin Hubble, Charles St. John, Albert Michelson, Albert Einstein, W. W. Campbell, and Walter Adams during a visit by Einstein to Mount Wilson. Hubble, a Carnegie Staff Member from 1919 until his death in 1953, with his colleague Humason discovered that the universe is expanding.

[49] *Zitterbewegung*
The literal translation is "tremble movement". It means jitter. Electrons do it. There's various explanations as to why, but see David Hestenes' 1990 paper *"The Zitterbewegung Interpretation of Quantum Mechanics".* Here's the

abstract: *The zitterbewegung is a local circulatory motion of the electron presumed to be the basis of the electron spin and magnetic moment. A reformulation of the Dirac theory shows that the zitterbewegung need not be attributed to interference between positive and negative energy states as originally proposed by Schroedinger. Rather, it provides a physical interpretation for the complex phase factor in the Dirac wave function generally. Moreover, it extends to a coherent physical interpretation of the entire Dirac theory, and it implies a zitterbewegung interpretation for the Schroedinger theory as well.* It was published in Foundations of Physics Vol 20, No. 10 (1990) pages 1213-1232. You can read it online at the Arizona State University *Geometric Calculus* website, see http://modelingnts.la.asu.edu/pdf /ZBW_I_QM.pdf.

[50] *Heuristic Viewpoint Concerning the Production and Transformation of Light*
Einstein's photoelectric paper appeared in the Annalen Der Physik in 1905, but there's a 1965 English translation in the American Journal of Physics at http:// dbserv.ihep.su/~elan/src/einstein05/eng.pdf. It starts thus: *It seems to me that the observations associated with blackbody radiation, fluorescence, the production of cathode rays by ultraviolet light, and other related phenomena connected with the emission or transformation of light are more readily understood if one assumes that the energy of light is discontinuously distributed in space. In accordance with the assumption to be considered here, the energy of a light ray spreading out from a point source is not continuously distributed over an increasing space but consists of a finite number of energy quanta which are localized at points in space, which move without dividing, and which can only be produced and absorbed as complete units.*

[51] *is thought to be ninety three billion light years across*
I got this from an interesting cosmology paper that put me straight about a few things. It's well worth a read. Get this: there are galaxies receding from us at more than c, and we can see them. Not all, just some. We can even see some galaxies where the light was emitted *when the galaxy was already receding from us at more than c.* Marvellous stuff. See *Davis, T. M.; Lineweaver, C. H.; "Expanding Confusion: common misconceptions of cosmological horizons and the superluminal expansion of the Universe", Publications of the Astronomical Society of Australia, 21, 97—109. URL: arXiv:astro-ph/0310808v2.*

[52] *156 billion light years is maybe more like it*
Take a look at the 2004 article by Robert Roy Britt interviewing Neil Cornish of Montana State University. It's called *Universe Measured: We're 156 Billion Light-years Wide!* You can find it on http://www.space.com/scienceastro-nomy/mystery_monday_040524.html. The full paper offers a slightly different slant on things. It's talking about a minimum figure, and it's getting bigger all the time. See *Neil J. Cornish, David N. Spergel, Glenn D. Starkman, Eiichiro Komatsu, "Constraining the Topology of the Universe", Phys.Rev.Lett. 92 (2004) 201302. URL: arXiv:astro-ph/0310233v1.*

[53] *Milgrom's MOND*
Mordehai Milgrom is the Israeli physics professor who came up with <u>M</u>odified <u>N</u>ewtonian <u>D</u>ynamics in the early eighties. See the wiki article <u>http://en.</u> <u>wikipedia.org/wiki/Modified_Newtonian_dynamics</u>, which mentions variants such as MSTG, TeVeS, and STVG, and links to the Lambda-CDM model, where you'll see the dark matter pie. The original paper is *A modification of the Newtonian dynamics as a possible alternative to the hidden mass hypothesis*, *Astrophysical Journal, Part 1 (ISSN 0004-637X), vol. 270, July 15, 1983, p. 365-370*. Also see the easy-reading August 2002 Scientific American article <u>http://www.astro.umd.edu/~ssm/mond/sad0802Milg6p.pdf</u>.

[54] *Pioneer Anomaly*
We all love a good mystery, and it would be a shame if we didn't have any left to solve. Sometimes I worry a little that life might become dull once we've cracked it, and the Pioneer anomaly is a cracker. See *Nieto, M. M.; Anderson, J. D.; "Seeking a solution of the Pioneer Anomaly", Fourth Meeting on CPT and Lorentz Symmetry, 8-11 Aug. 2007. URL: <u>arXiv:0709.1917v1</u>*. Also see *Saffari, R.; Rahvar, S.; "f(R) Gravity: From the Pioneer Anomaly to the Cosmic Acceleration". URL: <u>arXiv:0708.1482v1</u>*.

[55] *and then show me one atom of entropy*
This is from Sir Terry Pratchett's *Hogfather*, but I substituted "entropy" for "justice". There's some quite sharp stuff in his Discworld books.

[56] *Michael Faraday natural philosopher*
See the 1992 article by Jim Ealy and Louise Kemp of Princeton: <u>http://www.</u> <u>woodrow.org/teachers/chemistry/institutes/1992/Faraday.html</u>. It gives us two Faraday interviews which are a joy and pleasure to read.

[57] *Gravastar*
Gravastar is short for gravitational vacuum condensate star, and was proposed in 2001 by Pawel Mazur and Emil Mottola as an alternative to black hole theory. A gravastar is black, and it's actually more of a hole than a traditional black hole. See the Wikipedia article at <u>http://en.wikipedia.org/wiki/Gravastar</u> for an introduction. Note where it says: *This region is called a "gravitational vacuum", because it is a void in the fabric of space and time.*

[58] *Pace VanDevender's Irish bog*
You can find out more from a New Scientist article of 23rd December 2006 entitled *Black Holes in your backyard*. See <u>http://www.newscientist.com/chan</u> <u>nel/fundamentals/mg19225831.700-blackholes-in-your-backyard.html</u>. Here's a snippet: *So why would a GEA float? VanDevender believes the charged particles whizzing around the black hole generate a huge, oscillating magnetic field. As it neared the ground in a boggy region, the ball would induce currents in the Earth, setting up a magnetic field that would keep it above the ground, the same principle that levitates maglev trains.*

[59] *Ronald L Mallett trying to build a time machine*
See *Weak gravitational field of the electromagnetic radiation in a ring laser*, published in *Physics Letters A 269_2000, 214–217*. Also see web page <u>http://www.phys.uconn.edu/~mallett/Mallett2000.pdf</u>. Ron's nearby home page lists his interests, and black holes are top of the list. Also see *The Time Traveller: One Man's Mission To Make Time Travel A Reality* published by Doubleday in 2007.

[60] *Indeed, the fluctuations we see in the CMB*
Again from the 2004 article *Universe Measured: We're 156 Billion Light-years Wide!* by Robert Roy Britt interviewing Neil Cornish of Montana State University.

FIGURES

Figure 1 – Einstein at the patent office in Bern 1905
This is Einstein in his prime, sparkling and handsome, the real Einstein rather than the caricature. It was taken by his friend Lucien Chavan, an electrical engineer working for the Swiss post and telegraph office.

Figure 2 – Manchester University Maths tower, now demolished
From Nick Higham's *Mathematics Tower Photo Gallery* at http://www.maths. manchester.ac.uk/~higham/photos/mathstower/index.htm. Nick is the Richardson Professor of Applied Mathematics at the University of Manchester.

Figure 3 – Physics A-levels
This chart is from a CBI press release of August 2007: *CBI - Education and STEM Skills, Background Note for Journalists*. See http://www.cbi.org.uk/ and search on August 13th 2007.

Figure 4 – Top down and bottom up
Simple line drawing no image credit.

Figure 5 – Popular Science rocket science
I came across this on http://jetex.org/ which *serves the world-wide community of enthusiasts for models powered by micro rocket motors*. See http://jetex.org/ models/plans/plans-misc.html and *Daylight Skyrocket*. It's originally from *Popular Science* magazine and dates back to July 1954. See http://www.popsci.com/ where you can read: *Popular Science has been a leading source of science and technology news since its inception way back in 1872*. Here's a couple of Popular Science magazine covers. The future ain't what it used to be. But it will be.

Figure 6 – Stenography
From the English Wikipedia article *Stenotype*, see web page http://en.wikipedia
.org/wiki/Stenotype. Image by Gary D Robson.

Figure 7 – Quantum electrodynamics
Courtesy of Princeton University Press, this is the cover of the first edition of
Richard Feynman's book *QED: The Strange Theory of Light and Matter* published
in 1985. See their website at http://www.press.princeton.edu/. The history page at
http://press.princeton.edu/about_pup/puphist.html is interesting too, as is an essay by
Harold T Shapiro at http://press. princeton.edu/about_pup/shapessay.html.

Figure 8 – Feynman diagram
This drawing by me shows the virtual photons employed in quantum electrodynamics to account for electromagnetic attraction between a proton and an electron
in a hydrogen atom. The mathematics works, QED is a good theory, but people
forget that Feynman said nobody understood why it works. Hence they make the
mistake of assigning these virtual photons a degree of reality that isn't warranted.

Figure 9 – Challenger disaster
From the NASA website http://history.nasa.gov/sts511.html. Click on *image library*,
and it's on page 35. This is image STS51L-10178. A tragic loss.

Figure 10 – Woldemar Voigt, George FitzGerald, Hendrik Lorentz, Henri Poincaré
These public-domain images are from English Wikipedia articles, where they've
been copied from other sources. The Lorentz picture is his official portrait for the
1902 Nobel Prize, see http://nobelprize.org/nobel_prizes/physics/laureates/1902/
lorentz-bio.html. He shared it with Pieter Zeeman for the Zeeman Effect, or more
properly *in recognition of the extraordinary service they rendered by their
researches into the influence of magnetism upon radiation phenomena.*

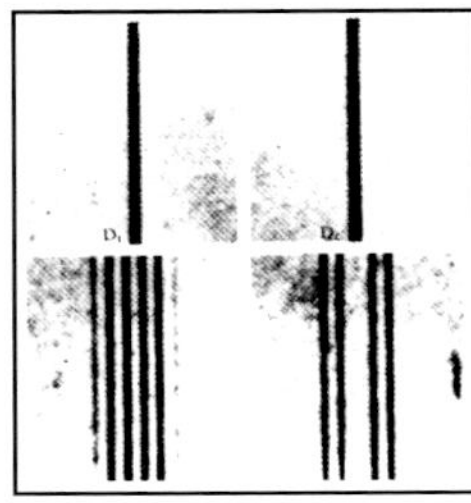

Figure 11 – Max Planck and Hermann Minkowski
Again these public-domain images are from English Wikipedia articles. Max Planck
won the Nobel Prize for physics in 1918 *in recognition of the services he rendered
to the advancement of physics by his discovery of energy quanta.* See his Nobel
Prize lecture at http://nobelprize.org/nobel_prizes/physics/laureates/1918/planck-
lecture.html. The Minkowski image is best viewed in a biography in the *MacTutor*

History of Mathematics Archive, see http://www-history.mcs.st-and.ac.uk/Biographies/Minkowski.html.

Figure 12 – Albert Einstein and Kurt Gödel
Image courtesy of Karin (Morgenstern) Papp. This photograph was taken on the 13th May 1947 by her father Oskar Morgenstern, IAS economics professor, one of the founders of game theory, and friend of Einstein and Gödel. My favourite story is how in December that year Morgenstern and Einstein were accompanying Gödel to his naturalization interview with the INS. Morgenstern later recounted: *When we came to Trenton, we were ushered into a big room, and while normally the witnesses are questioned separately from the candidate, because of Einstein's appearance, an exception was made and all three of us were invited to sit down together, Gödel in the center.* You can read more at the Institute for Advanced Study website http://www.ias.edu/, see page http://www.ias.edu/SpFeatures/kurt_godel/godel-2. html. Also see the 2005 New Yorker article *Time Bandits* by Jim Holt at http:// www.newyorker.com/archive/2005/02/28/050228crat_atlarge.

Figure 13 – Möbius strip
This image is from Gert van der Heijden's home page http://www.ucl.ac.uk/ ~ucesgvd/. Eugene Starostin and he wrote a paper in 2007 entitled *The shape of a Möbius strip,* which appeared in Nature Materials 6 on pages 563-567. You can find it online on Eugene's home page http://www.ucl.ac.uk/~ucesest/ and on Gert's home page as above where he says: *We study the mechanics of inextensible strips with applications to paper crumpling, fabric draping as well as general sheet processing. Geometrically this leads to the study of developable surfaces (surfaces flat in one direction). As part of this work we solved the long-standing problem of finding the shape of a Möbius strip.* Eugene and Gert are research fellows in the Centre for Nonlinear Dynamics of the Department of Civil, Environmental, & Geomatic Engineering at UCL.

Figure 14 – Ten pound note
From the English Wikipedia article *Sterling Banknotes* at http://en.wikipedia.org/ wiki/Sterling_banknotes. Permission granted by the Reproductions Officer at the Bank of England.

Figure 15 – Five dollar bill
From http://www.moneyfactory.gov/, the website of the US Bureau of Printing and Engraving within the US Treasury.

Figure 16 – Pennies
From *Don's World Coin Gallery* at http://worldcoingallery.com/ where you can find *over 25,000 coin photos from 450 states/countries.*

Figure 17 – Ledger
From Jenny Woolf's *Jabberwock* website http://www.jabberwock.co.uk/, courtesy of Barclays Group Archives. Jenny is the author of *Lewis Carroll In His Own Account*

and has used now-public bank records as a very useful source of biographical information. See the BBC feature at *http://news.bbc.co.uk/1/hi/magazine/6192829. stm*. Lewis Carroll aka the Reverend Charles Dodgson was of course the author of *Alice's Adventures in Wonderland* along with *The Hunting of the Snark*, and was also a mathematician and logician. See http://www.lewiscarroll.org/logic.html.

Figure 18 – Pirate
From the English Wikipedia article *Piracy*, see http://en.wikipedia.org/wiki/Piracy. This picture shows *Mic the Scallywag of the Pirates of Emerson Haunted Adventure Fremont CA.,* copyright © 2006 David Ball.

Figure 19 – Hell
This is the famous hell from the Hieronymus Bosch triptych *The Garden of Earthly Delights* painted around 1503. I was planning on showing an illustration of a negative carpet, but I couldn't find one. So this fits the bill just as well if you catch my drift. See http://en.wikipedia.org/wiki/Hieronymus_Bosch for more.

Figure 20 – Parallel worlds
This is a trio of "blue marble" images from NASA, see the *Visible Earth* website, page http://visibleearth.nasa.gov/view_rec.php?id=2429 where the credit says: *This spectacular "blue marble" image is the most detailed true-color image of the entire Earth to date. Using a collection of satellite-based observations, scientists and visualizers stitched together months of observations of the land surface, oceans, sea ice, and clouds into a seamless, true-color mosaic of every square kilometer of our planet... Much of the information contained in this image came from a single remote-sensing device - NASA's Moderate Resolution Imaging Spectroradiometer.*

Figure 21 – Intelligent Design
This excellent cartoon by Tony Auth of the Philadelphia Inquirer is all over the Internet because it is such a cracker. See http://www.linesandcolors.com/2006 /10/page/2/ for a potted history with a link to an archive at http://www.gocomics .com/tonyauth. Copyright AUTH © 2005 The Philadelphia Inquirer, reprinted with permission of UPS, all rights reserved.

Figure 22 – Checker-shadow illusion by Edward Adelson
Edward H. Adelson is John and Dorothy Wilson Professor of Vision Science at the Department of Brain and Cognitive Sciences at MIT. See his homepage at http://web.mit.edu/persci/people/adelson/index.html for both the original checker-shadow illusion and the explanation: *The visual system uses several tricks to determine where the shadows are and how to compensate for them, in order to determine the shade of gray "paint" that belongs to the surface. The first trick is based on local contrast. In shadow or not, a check that is lighter than its neighboring checks is probably lighter than average, and vice versa...*

Figure 23 – Checker-shadow illusion revealed
As above. All I did was slap a patch of grey on there. Try it yourself.

Figure 24 – Explain it to your grandmother
Courtesy of James Charles, artist. See his Masterwork Gallery at http://www. masterworkgallery.com/, click on *Family Portrait Painting* at the bottom then *Grandparent Portrait Painting*, and there she is, chuckling and purling away.

Figure 25 – Red balloon, red bus, and red red ruby
The balloon is from the Big Red Balloon Company at http://www.bigredballoon .com/ who do sightseeing trips in Florida. The bus is from Timebus Travel at http://www.timebus.co.uk/ who provide traditional buses for hire within London and the Home Counties. The ruby is from *Ruby and Sapphire* by Richard W Hughes, see http://www.ruby-sapphire.com/brilliance_windows_extinction.htm.

Figure 26 – Cannonball in space
Image copyright © David A. Hardy / www.astroart.org. It's really *The Millennium Planet*, see http://www.hardyart.demon.co.uk/pages-gallery/p-extra1.html where Dave says: *The Millennium Planet depicts a planet of the star Tau Boötis - a huge, bluish gas giant, bigger than Jupiter. At the time, this was thought to be the first visual confirmation of such a world. I postulated a close moon, which would be tidally disrupted, giving rise to tectonic and volcanic features, hence the glows on its night side.* Dave tweaked the image a little to make the moon look more like a cannonball for me. By the by, I first saw this image on the Particle Physics and Astronomy Research Council website http://www.pparc.ac.uk. This now redirects to http://www.scitech.ac.uk/ because PPARC merged with the Council for the Central Laboratory of the Research Councils (CCLRC) on 1 April 2007 to form the Science and Technology Facilities Council (STFC), who administer research funding. Like for Jodrell Bank.

Figure 27 – Gravity spaceship
This is the name you find on the image on the internet, but actually it's GOCE. That's ESA's *Gravity field and steady-state Ocean Circulation Explorer developed to bring about a whole new level of understanding of one of the Earth's most fundamental forces of nature – the gravity field.* See http://www.esa.int/esaLP/LP goce.html for details.

Figure 28 – Myosin structure
From the Laboratory of Molecular Cardiology (LMC), National Heart Lung and Blood Institute, Bethesda, and in particular from a James R Sellers article re research activity in the Cellular and Molecular Motility Section. See web page http://dir. nhlbi.nih.gov/labs/lmp/myosinlab.asp.

Figure 29 – Elastin
Courtesy of Garland Science Textbooks, this image is figure 19-52 from *Molecular Biology of the Cell, Fourth Edition*, copyright © 2002 Bruce Alberts, Alexander Johnson, Julian Lewis, Martin Raff, Keith Roberts, and Peter Walter. See http:// www.garlandscience.co.uk/textbooks/0815332181.asp.

Figure 30 – Hydrogen fusion
See the English Wikipedia article *Proton-proton chain reaction*, web page http://en .wikipedia.org/wiki/Proton-proton_chain_reaction. Original image by Borb, revised and reproduced under the terms of the GNU Free Documentation License.

Figure 31 – Annihilation schematic
This is based on an image on the American Physical Society *Physics Central* website, see the "Antihydrogen Antics" article at http://www.physicscentral.com/ explore/action/antics-1.cfm. Here's how it starts: *As Star Trek fans know well, the fuel for warp drive is antimatter. No science fiction stuff this - antimatter was predicted by Paul Dirac in 1928. At the time, the only known elementary particles were the positively charged proton and the negatively charged electron. Dirac combined the recently discovered quantum physics with special relativity and produced an equation that seemed to predict two kinds of particles: those with positive energies, as expected, and those with negative energies, which didn't make any sense. After pondering this result for several years, Dirac realized that these negative energies actually correspond to "antiparticles" of positive energy with the same mass as protons and electrons but with opposite charge.*

Figure 32 – Annihilation animation
From the Goddard Space Flight Center Scientific Visualization Studio, see http://svs .gsfc.nasa.gov/goto?182, lead animator Edgar Russell, scientist David Leisawitz.

Figure 33 – Slinky
Courtesy of DK Images, the online image encyclopaedia: *3.5 million images, 70 thousand artworks, 180 talented photographers, 1 experienced sales team.* See http://www.dkimages.com and search on slinky.

Figure 34 – Space
This is an image of the M81 galaxy from the Spitzer website http://www.spitzer
.caltech.edu/spitzer/, courtesy NASA JPL/Caltech. See page http://gallery.spitzer.
caltech.edu/Imagegallery/image.php?image_name=sig07-009 where the credit reads:
*Hubble data: NASA, ESA, and A. Zezas (Harvard-Smithsonian Center for
Astrophysics); GALEX data: NASA, JPL-Caltech, GALEX Team, J. Huchra et al.
(Harvard-Smithsonian Center for Astrophysics); Spitzer data: NASA/JPL/Caltech/S.
Willner (Harvard-Smithsonian Center for Astrophysics).* It's a composite created
using visible-light data from the Hubble space telescope, infra-red from the Spitzer
space telescope, and ultraviolet from the Galaxy Evolution Explorer space telescope.
Spitzer is a NASA mission run by the Jet Propulsion Laboratory (JPL) in Pasadena,
see http://www.jpl.nasa.gov/index.cfm. The Jet Propulsion Laboratory is part of
NASA but is staffed and run by Caltech in Pasadena. This is Spitzer:

*SIRTF points its high-gain antenna
towards the Earth for downlinking
recent observations and uplinking
new observing instructions. Credit:
NASA/JPL-Caltech.*

Figure 35 - Fist
From http://www.livepencil.com/ by Ramon Gonzalez Teja, freelance illustrator and
graphic designer. Ramon was born in Dallas and lives in Madrid. The fist is one of
his wallpapers.

Figure 36 – Energy/mass conversion
Image courtesy of Professor Wolfram von Oertzen, leader of the working group on
nuclear instrumentation at the Hahn Meitner Institute in Berlin, which investigates
the structures of matter and materials and manufacturing techniques for photovoltaic
cells. See http://www.hmi.de/ and in particular the Public Awareness of Nuclear
Science (PANS) section and web page http://www.hmi.de/bereiche/SF/SF7/PANS/
english/dualismus/dual_frame.html. Click on *Generation of particles*, where the
intro to this image reads: *one of the consequences of the equivalence of energy and
mass is that in a collision of two particles, occurring at a very high energy E, new
electromagnetic waves (radiation), as well as new particles can be produced.*

Figure 37 – Gyroscope
From http://www.physlink.com/. Physlink is a physics and astronomy education, research, and reference web site, and a forum for professionals, students and others. The e-store now links to http://www.xump.com/ but the *Space Wonder Gyroscope* is still there. Talk about apt. You couldn't make it up.

Figure 38 – Wave in the surf
This is from surfer-dude website http://www.surf-costarica.com/. Costa Rica does look good. See http://www.surf-costarica.com/images/large/sd-puertoviejo-cocles2-5-1-7.jpg.

Figure 39 – Compton scattering
From *HyperPhysics*, copyright © 2006 C. R. Nave. Carl "Rod" Nave is Associate Professor at the Department of Physics and Astronomy at Georgia State University in Atlanta. See http://hyperphysics.phy-astr.gsu.edu/hbase/quantum/compton.html for details. HyperPhysics is a really good website chock full of very clear information laid out in a very accessible form. A must.

Figure 40 - Pair production
From the English Wikipedia article *Pair production*, see http://en.wikipedia.org/wiki/Pair_production. Image created by David Horman, reproduced under the terms of the GNU Free Documentation License (GNU FDL).

Figure 41 – Toroidal photon electron
This is from *Is the electron a photon with toroidal topology?* and is reproduced with the permission of John G Williamson and Martin B van der Mark. This paper was really important in terms of my understanding. One day perhaps we'll take a cold hard look at why it took six years to make it into a journal, and why it languished thereafter. You can view it online at http://www.cybsoc.org/electron.pdf.

Figure 42 – Balloon knot
Original photograph by Martin Braun from Bad Vöslau in Austria. It's a spa and wine town about twenty miles south of Vienna.

Figure 43 – PET scan
From the English Wikipedia article *Positron Emission Tomography*, see http://en.wikipedia.org/wiki/Positron_emission_tomography. Image by Jens Langner.

Figure 44 – Paper-strip circle with a twist, and the same strip in figure-8 form
Simple line drawing no image credit.

Figure 45 – Helix
Helix tube image produced using the mathematical visualization software 3D-XplorMath. See http://3d-xplormath.org/, and follow the link to http://3d-xplormath.org/consortium.html telling you about the consortium of mathematicians including Chuu-Lian Terng, David Eck, Dick Palais, Hermann Karcher, Markus Schmies, Martin Guest, Dan Steinberg, Matthias Weber, Michael Murray, Patrick Iglesias, Takashi Sakai, Traudel Karcher. Xah Lee, Luc Benard, Bob Palais, Nam Trang, and Gale Paeper. The helix tube image itself is at http://virtualmathmuseum.org/SpaceCurKves/helix/helix.html.

Figure 46 – Knots
This is a crop from *A Knot Zoo*, an image created by Doctor Robert Scharein, freelance developer of software for mathematical and scientific visualization. See *The Knotplot Site* at http://knotplot.com/ and the image itself at http://knotplot.com/zoo/. It was created with *KnotPlot*, a program to visualize and manipulate mathematical knots in three and four dimensions. Rob has a degree and masters in physics and a doctorate in computer science.

Figure 47 – Tesla coil
From the *Tesla Memorial Society of New York* website, see http://www.teslasociety.com/ and in particular page http://teslasociety.com/lasvegas.htm. Nikola Tesla really was a genius, *the genius who lit the world*. Personally I'm pretty sure he knew a hundred years ago that AC was fundamental rather than DC, which is some going. Whilst he's sometimes thought of as being eccentric later in life, check out the society board members and you'll see how seriously he's taken. And check out what you can of his dynamical theory of gravity on http://netowne.com/technology/important/. He wasn't far off.

Figure 48 – Gamma radiation
Actually this is picture of the star Sirius, but you get the idea. There's an awful lot of energy tied up in something as small as a battery. The image is from the HubbleSite, see http://hubblesite.org/ and in particular http://hubblesite.org/newscenter/archive/releases/2005/36/image/a/. Image credit: NASA, H.E. Bond and E. Nelan (Space Telescope Science Institute, Baltimore, Md.); M. Barstow and M. Burleigh (University of Leicester, U.K.); and J.B. Holberg (University of Arizona).

Figure 49 – Road mirage and bubble
Both these pictures were taken by Mila Zinkova, and can be found on Wikipedia articles http://en.wikipedia.org/wiki/Mirage and http://en.wikipedia.org/wiki/Soap_

<u>bubble</u>. Also see her gallery at <u>http://home.comcast.net/~milazinkova/Fogshadow</u> <u>.html</u> where you can see a host of other beautiful and interesting pictures.

Figure 50 – Falaco soliton
From *Kiehn, R. M. (2004-2nd edition 2007) Cosmology, Falaco Solitons and the Arrow of Time, Non-Equilibrium Systems and Irreversible Processes Vol 2, Lulu Enterprises, Inc., 3131 RDU Center, Suite 210, Morrisville, NC 27560.* See <u>http://www.lulu.com/kiehn</u>:

The fact that the oblique projection yields a circle (and therefore is conformal) is a clue that the surface dimple is related to a minimal surface. The actual surface dimple depression is about 2 mm (but is highly contrast-distorted in the photo). The black spots on the bottom of the pool are about 10 cm in diameter. The pool depth is about 30 cm.

Figure 51 – Toroidal photon electron and mirror image positron
The electron is again from *Is the electron a photon with toroidal topology?* by John Williamson and Martin van der Mark. I flipped the electron to make the positron.

Figure 52 – Twisted cube
Image produced by Steve Hill using the modelling software AC3D, see <u>http://website.lineone.net/~steve_hill/Graphics/AC3D/AC3DResources.html</u>. Steve is also into astronomy, and has some great pictures on his Astrobrowser at <u>http://myweb.tiscali.co.uk/blobchaser/Astronomy/Navigator/TheNavigator.html</u>.

Figure 53 – Refraction
From the *Arizona Collaborative for Excellence in Preparation of Teachers* website, part of the *Patterns in Nature* project-oriented laboratory science course prepared by Professor James Walter Mayer of Arizona State University. See http://acept.la.asu.edu/PiN/info/patt.html page http://acept.la.asu.edu/PiN/rdg/refraction/refraction.shtml.

Figure 54 – Pinwheel
From http://www.jbum.com/pixmagic/galspirals.html. Jim Bumgardner is a software technologist, teacher, and devotee of recreational programming. Pixel Magic is one of his image generation programs, which he describes as *an exploratory graphics program for the mathematically literate.*

Figure 55 – Faraday effect
From the English Wikipedia article *Faraday Effect*, see http://en.wikipedia.org/wiki/Faraday_Effect. Image by "DrBob" Bob Mellish, reproduced under the terms of the GNU FDL.

Figure 56 – YBCO superconductor
Image courtesy of Marcus Hewat who produced it using his 3D VRML structure drawing application xtal-3d. I first saw it on the Laue-Langevin Institute for neutron science and technology website http://www.ill.fr/, see web page http://www.ill.eu/science-technology/science-at-ill/magnetism/perovskites/.

Figure 57 – Meissner effect: levitation of a magnet above a superconductor
From the English Wikipedia article *Meissner Effect,* see http://en.wikipedia.org/wiki/Meissner_effect. Image by Mai-Linh Doan reproduced under the terms of the GNU FDL.

Figure 58 – Reference frames colour your vision
This is baby Rhys, born on February 9th 2008. He's about six months old here.

Figure 59 – Zero gravity
This image is from the Zero Gravity Organisation, see http://www.gozerog.com/ web page http://www.gozerog.com/images/imggal/30_a.jpg. Zero G was co-founded by space visionary and entrepreneur Dr. Peter H. Diamandis, veteran astronaut Dr. Byron K. Lichtenberg, and NASA engineer Ray Cronise. They fly a Boeing 727 in a rollercoaster parabolic arc going from 34,000 to 24,000ft. The picture title is *Fernmarie in 0g*. Fernmarie Rodriguez from Arecibo in Puerto Rico was a NASA intern and attended the ISU, and is building a fine aerospace track record.

Figure 60 – Principle of equivalence
From Nick Strobel's website http://www.astronomynotes.com/, see web page http://www.astronomynotes.com/relativity/s3.htm. Nick is Professor of astronomy at Bakersfield College and director of the William M Thomas Planetarium. His website is laden with a wealth of interesting and useful information such as this:

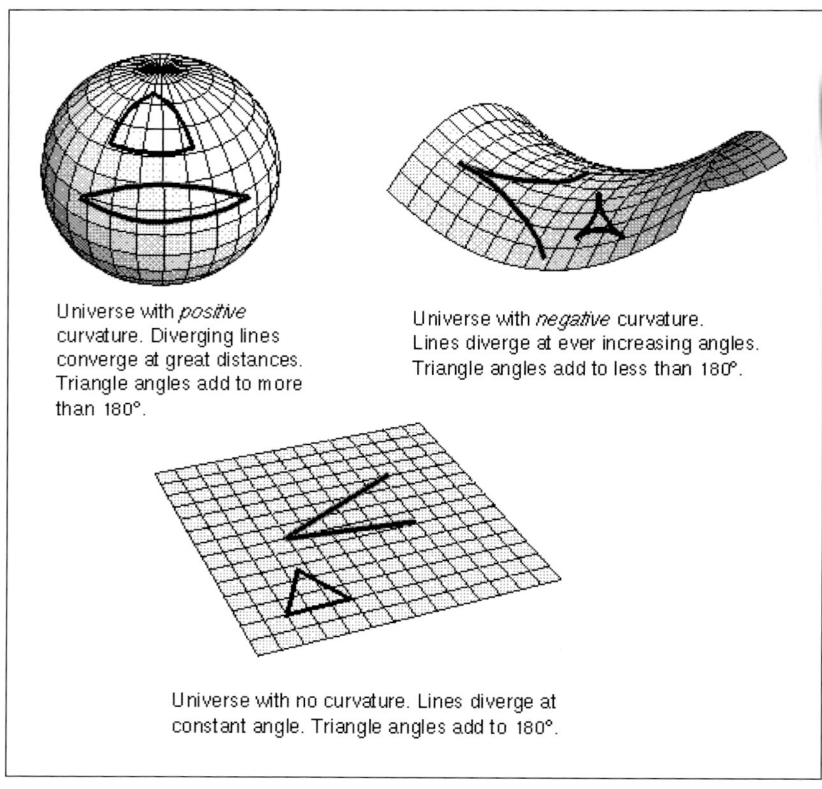

Figure 61 – Pound-Rebka experiment
This is another one from the *HyperPhysics* website by Rod Nave, associate professor of physics at Georgia State University, Atlanta. See http://hyperphysics.phy-astr.gsu.edu/hbase/hframe.html, then click on *Relativity* followed by *Gravitational Time Dilation* and there you'll see *The Harvard Tower Experiment* and this image.

Figure 62 – Crystal spheres
This is *The Flammarion Woodcut*, see http://en.wikipedia.org/wiki/Flammarion_woodcut. It's a wood engraving that first appeared in a book on meteorology by Camille Flammarion in Paris in 1888. It depicts a breakthrough moment when a pilgrim on a journey pokes his head through the crystal-sphere sky to see the workings of the universe beyond. The title reads: *a medieval missionary recounts that he has found the point where the sky and the Earth meet*. Colour added by Tobin James Mueller for the ArtsForge *Discovering the Heavens* collaborative art project, see http://www.tobinmueller.com and http://www.artsforge.com/visionforge/discov.html.

Figure 63 – Brightness illusion by R Beau Lotto
From *lottolab,* see http://www.lottolab.org/. Lottolab is headed up by Dr Ray Beau Lotto, lecturer in neuroscience within the Institute of Ophthalmology at UCL. Research themes are *human perception and processing, insect visual development processing and behaviour, artificial life systems, and art science and education installations.* See http://www.lottolab.org/Brightness illusions page.html#. Strangely enough, once you get the hang of these colour illusions, they don't fool you any more. You learn to see what's there. Apart from the Big Spanish Castle, which really blows you away. See http://www.johnsadowski.com/big_spanish_castle.php.

Figure 64 – Bat sonar
Drawn by me via a little public-domain pasting. It's akin to a low-res image in *Science News.* See http://www.sciencenews.org/ and the May 14th 2005 cover story *Learning to Listen* by Sid Perkins at http://www.phschool.com/science/science _news/articles/learning_to_listen.html. The narrative there reads: *HOMING IN. The distance between pressure pulses in a bat's ultrasonic chirps and echoes (represented by white and red wave outlines respectively) determines how small a target the predator can detect. (M Tuttle/Bat Conservation Intl./S.Norcross).*

Figure 65 – Heat as molecular motion
Simple line drawing no image credit. See http://en.wikipedia.org/wiki/Kinetic _theory for more, including a GIF moving image. Sit and watch it for a while. Then remind yourself that things move through space, not through time.

Figure 66 – Spin flip
I couldn't find a suitable picture online to show 3-dimensional spin, so this is another one by me. I was going to draw billiard balls, but I thought pool balls would be better. The 8-ball represents an electron, and I added an extra loop to turn the 8 into a trefoil for the central ball representing the proton. There ought to be an extra loop to the proton spin but it started looking too fussy. See http://en.wikipedia.org /wiki/21_cm_line for background information on the hyperfine transition, along with http://en.wikipedia.org/wiki/Spin_(physics).

Figure 67 – Beans in a bucket
Courtesy of *High Country Coffee.* They sell coffee, and you should be smelling it by now. See http://highcountrycoffee.com/, click on *Shop* then *Papua New Guinea coffee* and there it is, the bucket of beans that shot down the arrow of time. These are of course coffee cherries. The beans are inside. But they're not really beans.

Figure 68 – Clocks don't really run
I couldn't track down the running clock image I planned to use, so this is drawn by me. Can I add that athletes run, horses run, and I suppose at a pinch you could say cars run because they do actually travel. But engines don't really run. Nor do taps, nor do noses, and nor do clocks. Of course, the many and varied ways we use the word "run" doesn't help at all. See http://www.answers.com/topic/run. I counted sixty four.

Figure 69 – Molecules
I wanted a degree of consistency across various figures, so I've drawn this one from scratch. However it's based on an image on page 381 of *College Physics*, 5th edition, by Jerry D Wilson & Anthony Buffa, published by Prentice Hall. *College Physics is* a thousand-page treasure trove. See http://www.amazon.com/College-Physics-Jerry-D-Wilson/dp/0130676446.

Figure 70 – Time dilation is based on Pythagoras' Theorem
From the English Wikipedia article *Time Dilation* by Alfio and DVdm and many others. See http://en.wikipedia.org/wiki/Time_dilation. Image by mdd4696.

Figure 71 – Clocks clock up motion, not time
Image by Jonathan Kotta, from Wikipedia commons. It shows the interior parts of a Hamilton pocket watch. Dual-licensed under the GFDL and CC-By-SA-2.5, 2.0, and 1.0. See http://commons.wikimedia.org/wiki/Image:Hamilton_926_movement.jpg.

Figure 72 – Rubber sheet analogy for gravity
Artwork copyright © 2005 Boris Starosta, see http://www.starosta.com. I saw it on the Meta Research website http://metaresearch.org/home.asp, see page http://meta research.org/cosmology/PhysicsHasItsPrinciples.asp. And I quote: *Meta Research was founded in 1991 in response to the broad problem of getting support to do research on promising but unpopular alternative ideas in astronomy that have trouble getting funded from the usual sources.*

Figure 73 – Knot in a sheet
This is a picture I took at home on my digital camera. It's a piece of cling film with an overhand knot in it, stretched over a small picture frame.

Figure 74 – Piezoelectric effect
Image by Mael Guennou, from Wikimedia Commons (http://commons.wikimedia. org/wiki/Image:SchemaPiezo.gif) released under GFDL (http://commons.wikimedia. org/wiki/Commons:GNU_Free_Documentation_License) and CC-BY-SA (http:// creativecommons.org/licenses/by/3.0/). You can see it in action in the Wikipedia article *Piezoelectricity*, see http://en.wikipedia.org/wiki/Piezoelectric. Note the link to http://en.wikipedia.org/wiki/Piezoresistive_effect.

Figure 75 – Alternating current
An image similar to this was on the BBC GCSE "bitesize" website, but has now gone due to dumbing down. So I've had to draw it myself. See http://www. bbc.co.uk/schools/gcsebitesize/science/.

Figure 76 – Gravitational lensing
From the HubbleSite newscenter release entitled *Hubble finds double Einstein ring*. See http://hubblesite.org/newscenter/archive/releases/exotic/gravitational%20lens/20 08/04/full/ for details such as: *This is an image of gravitational lens system SDSSJ0946+1006 as photographed by Hubble Space Telescope's Advanced Camera*

for Surveys. The gravitational field of an elliptical galaxy warps the light of two galaxies exactly behind it. The massive foreground galaxy is almost perfectly aligned in the sky with two background galaxies at different distances. Credit: NASA, ESA, and R. Gavazzi and T. Treu (University of California, Santa Barbara), and the SLACS team. There's more, like this "five star" gravitational lens:

Figure 77 – Shapiro effect
I spotted this in Science Daily, *one of the Internet's leading online magazines and Web portals devoted to science, technology, and medicine.* See http://www.science daily.com/releases/2006/09/060914094623.htm. It's an article from September 2006 entitled *General Relativity Survives Gruelling Pulsar Test: Einstein At Least 99.95 Percent Right*. It concerns an international team led by Professor Michael Kramer of Jodrell Bank and three years' observations of double pulsar PSR J0737-3039A/B. The image depicts the double pulsar system being tracked by the radio astronomers who discovered it, including Dr. Duncan Lorimer and Dr. Maura McLaughlin of the University of Manchester now at West Virginia University: *The pulsars are the remnants of two massive stars that burned out by way of supernova explosions. They measure just 12 miles across, but each weighs more than our own Sun. Note the "bend" in the space-time fabric from the sheer mass of the two bodies.* See http://www.jodrellbank.manchester.ac.uk/news/2006/einstein/ for more.

Figure 78 – Plastic lens used in astronomical microlensing simulation
From the MOA website http://www.phys.canterbury.ac.nz/moa/demonstration.html which describes a demonstration of gravitational microlensing effects using plastic lenses to mimic the gravitational field. The home page at http://www.phys. canterbury.ac.nz/moa/ tells you more, saying: *MOA (Microlensing Observations in Astrophysics) is a Japan/NZ collaboration that makes observations on dark matter,*

extra-solar planets and stellar atmospheres using the gravitational micro lensing technique at the Mt John Observatory in New Zealand. Credits: Lenses David Cochrane, IRL Ltd, Wellington; Photography, Kevin Taylor and Philip Yock University of Auckland; File Compression, Amstore Ltd, Auckland.

Figure 79 – Photon strip and figure-of-eight electron strip
Again from *Is the electron a photon with toroidal topology?* by John Williamson and Martin van der Mark. The underlying text gives the following description:

a) Twisted strip model for one wavelength of a photon with circular polarisation in flat space. The B-field is in the plane of the strip and the E-field is perpendicular to it.

b) A similar photon in a closed path in curved space with periodic boundary conditions of length λ_c. The E-field vector is radial and directed inwards, and the B-field is vertical. The magnetic moment μ, angular momentum L, and direction of propagation with velocity c are also indicated.

Figure 80 – Satellite time dilation
From the English Wikipedia article *Global Positioning System*, see web page http://en.wikipedia.org/wiki/Global_Positioning_System. Image created by Phil Fraundorf, associate professor of physics and astronomy at the University of Missouri in St Louis. See his website at http://www.umsl.edu/~fraundorfp/.

Figure 81 – Newton was right on the money
This picture of Newton on a series D pound note is from the Bank of England website http://www.bankofengland.co.uk/, permission granted by the Reproductions Officer at the Bank of England. You can also see Newton on a note on Ulf Leonhardt's Invisibility page at http://www.st-andrews.ac.uk/~ulf/invisibility.html. Ulf is chair of Theoretical Physics at the University of St Andrews, and tells us: *Astronomy students might object that there is something distinctly wrong with the celestial ellipses on the one-pound note: the Sun is supposed to sit at the focal point, not the centre! Instead of Newton's ellipses, the Bank of England plotted the ellipses of Newton's rival, Robert Hooke.*

Figure 82 – There is no free lunch
Original photograph by Martin Braun from Bad Vöslau in Austria.

Figure 83 – LIGO
From http://www.ligo-la.caltech.edu/ and in particular from http://www.ligo-la.caltech.edu/worksheets/SEC/LIGO101.pdf where we read: *LIGO is a scientific collaboration of the California Institute of Technology (Caltech) and the Massachusetts Institute of Technology (MIT). Funded by the National Science Foundation.*

Figure 84 – The world is painted in light
Copyright © Neil Walker, Hyberian Digital Art, see http://hyberian.com/. He's an artist and photographer who works with digital images and photomontage, and his strapline is *painted in light*. This image is part of his Blue Orion Nebula. Neil also did the cover.

Figure 85 – Brilliant one-trick pony
From the Lexington-Fayette Urban County Arts Review Board web page http://www.lfucg.com/affiliated/arts_review/, featuring "Circus Horse" by Judy Anderson. Judy is an artist in Columbus Ohio, see http://artbyanderson.com/, and this is actually a fibreglass horse. The event was *Horse Mania*, a public art project sponsored by LexArts (formerly the Lexington Arts and Cultural Council) where 78 artists decorated lifesize fibreglass horses that were displayed throughout Lexington, Kentucky.

Figure 86 – Non-spinning black hole
This is from the Harvard-Smithsonian Centre for Astrophysics, see the November 2006 press release *Spinning Black Hole Pushes the Limit* at http://www.cfa.harvard .edu/press/2006/pr200630.html. A team led by Jeffrey McClintock and Ramesh Narayan used NASA's Rossi X-ray Timing Explorer satellite data to estimate the spin of black hole GRS1915+105 at 950 revolutions per second. The image itself was produced by Melissa Weiss of the Chandra X-ray Center. The CXC is operated for NASA by the Smithsonian Astrophysical Observatory, see http://chandra. harvard.edu/. Whilst I have some issues with black hole spin, here's the associated image of a spinning black hole for balance:

Figure 87 – Projection
From the *How Stuff Works* article *How Projection Television Works* by Tracy V. Wilson and Craig Freudenrich, see http://electronics.howstuffworks.com/projection-tv.htm/printable. Image courtesy of Philips Research, where Martin van der Mark hails from, see http://www.research.philips.com/.

Figure 88 – Light
This is from the PIRA website, see http://physicslearning.colorado.edu/PiraHome/. PIRA is the Physics Instructional Resource Association, and is: *an association of*

professionals dedicated to the support and advancement of physics education. We work together and in alliance with physics educators and support specialists to develop effective teaching tools and techniques to aid and promote physics education. PIRA members include teachers, professors, instructional laboratory specialists, lecture demonstration specialists, community outreach coordinators and others. We support physics education in elementary schools, high schools, community colleges, and universities. Although most members reside in the United States, PIRA is a global association. Their website includes a large collection of drawings that are free for educational use. All the images were created by Machele Cable of Wake Forest University, with input from Aaron Titus of High Point University and Cary Busby of Arbor Scientific. In addition there's the PIRA Demonstration Bibliography at http://physicslearning.colorado.edu/Pira.asp. This contains 9265 lecture demonstration entries ranging from mechanics to waves to thermodynamics, electricity and magnetism, optics, and much more. Commendable.

Figure 89 – Circular polarization
This is two images combined, both from Rod Nave's *HyperPhysics* at Georgia State University. See his comprehensive section on polarization at http://hyperphysics .phy-astr.gsu.edu/Hbase/phyopt/polarcon.html including the quarter-wave plate used to produce circular polarization, and circular polarization itself.

Figure 90 – Polarizing filters
Courtesy of Professor Steve Dutch, Natural and Applied Sciences, University of Wisconsin Green Bay. See his personal website http://www.uwgb.edu/DutchS/, and web page http://www.uwgb.edu/DutchS/petrolgy/genlight.htm.

Figure 91 – Swell waves
From the English Wikipedia article *Swell (ocean)*, see web page http://en.wikipedia .org/wiki/Swell_(ocean). Image by Phillip Capper. It's *Easterly swell, Lyttelton Harbour, New Zealand, 29 July 2008.*

Figure 92 – Elastic band
Courtesy of *Practical Physics*, a joint project of the Nuffield Curriculum Centre and The Institute of Physics, providing a website for physics teachers to share skills and experience re experiments. It's particularly hands-on, and delivers real grasp. See http://www.practicalphysics.org/. The image is from *Stretching and compressing materials* at http://www.practicalphysics.org/go/Experiment_188.html.

Figure 93 – Isaac Newton
From the English Wikipedia article http://en.wikipedia.org/wiki/Isaac_Newton, which advises that this is *Godfrey Kneller's 1689 portrait of Isaac Newton aged 46.*

Figure 94 – Michael Faraday, James Clerk Maxwell, and Hermann Weyl
The Faraday and Maxwell images are from English Wikipedia articles http://en. wikipedia.org/wiki/Michael_Faraday and James_Clerk_Maxwell. The Weyl picture is just public domain, but see William Straub's website http://www.weylmann.com/.

Figure 95 – Structure within the atom
Image courtesy of The Contemporary Physics Education Project, see http://www.cpepweb.org/ where we learn: *The Contemporary Physics Education Project is a non-profit organization of teachers, educators, and physicists located around the world. CPEP materials present the current understanding of the fundamental nature of matter and energy, incorporating the major research findings of recent years. During the last ten years, CPEP has distributed more than 200,000 copies of its charts and other products.* This is the main chart:

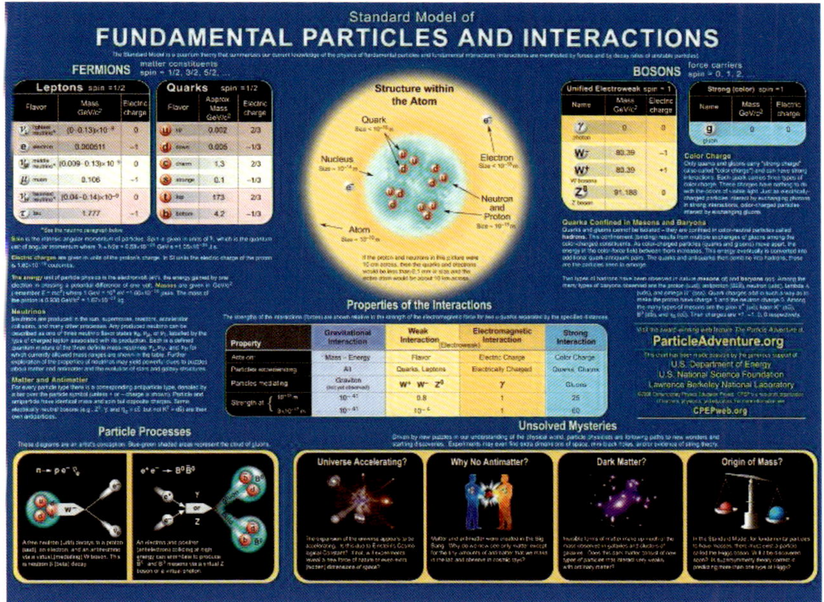

Figure 96 – Strip electron and möbius doughnut electron
Again from *Is the electron a photon with toroidal topology?* by John Williamson and Martin van der Mark.

Figure 97 – Light wave
From the English Wikipedia article *Light Wave*, see http://en.wikipedia.org/wiki/Light_wave. Image created by Heron, reproduced under the terms of the GNU FDL. The caption reads: *A linearly-polarized light wave frozen in time and showing the two oscillating components of light; an electric field and a magnetic field perpendicular to each other and to the direction of motion (a transverse wave).*

Figure 98 – Paper strip
Simple line drawing no image credit.

Figure 99   – Photon lattice
Figure 100 – 1022 keV photon and 511 keV photon
Figure 101 – Photon splitting, and photon undergoing pair production
All simple line drawings, no image credit.

Figure 102 – De Broglie atom
Courtesy of Kenneth Snelson, whose art *is concerned with nature in its primary aspect, the patterns of physical forces in three dimensional space.* See his website at http://www.kennethsnelson.net/ and in particular page http://www.kennethsnelson .net/circ_sph/index.htm.

Figure 103 – Solvay congress 1927
Photograph by Benjamin Couprie, Institut International de Physique Solvay, courtesy AIP Emilio Segre Visual Archives. It's from the American Institute of Physics website at http://www.aip.org/, see web page http://www.aip.org/history /einstein /quantum1.htm. They have a *Center for History of Physics* with a variety of interesting exhibits, including http://www.aip.org/history/einstein/. Also see wiki page http://en.wikipedia.org/wiki/Solvay_Conference, where you can follow the links to find out more about the participants.

Figure 104 – The electromagnetic spectrum
From the English Wikipedia article *Electromagnetic Spectrum*, see http://en.wiki pedia.org/wiki/Electromagnetic_spectrum. Original source is MYNASA, *Mentoring and inquirY using NASA Data for Atmospheric and Earth science for Teachers and Amateurs*, see web page http://mynasadata.larc.nasa.gov/glossary.php?&letter=E.

Figure 105 – The particle zoo
Image courtesy of Julie Peasley of *The Particle Zoo*. Julie sells these excellent soft toys as an aide memoire and to promote particle physics. The more massive particles are filled with gravel to give them extra weight. See http://www.particlezoo.net/.

Figure 106 – Trefoil knot
Drawn by me. It's based on the square-section trefoil in Jaap Zonneveld's *Stereoscopic Show of Mathematical Models*, which you can see at http://home .iae.nl/users/zonneve/.

Figure 107 – Photon bullwhip and neutrino bullwhip
The photon bullwhip is from Walt Schaaf's *Schaaf Saddlery and Leatherwork*, Cincinnati, see http://www.schaafleatherwork.com/pages/other/bullwhip.html. The neutrino bullwhip is from Tim Weske, swordmaster and trainer to the stars, see http: //www.timweske.com/weapons.shtml.

Figure 108 – B-sub meson oscillates between particle and antiparticle
This is another paper-strip analogy drawn by me. I reiterate that we're really dealing with spacewarp here, not some flat substance like paper.

Figure 109 – Kissing in 2 dimensions
From the English Wikipedia article *Kissing Number problem*, web page http://en.wikipedia.org/wiki/Kissing_number_problem. Image by N Mori.

Figure 110 – Kissing in 3 dimensions
From *Newton and the kissing problem* by George Szpiro as featured in *Plus* magazine, see http://plus.maths.org. For information: *Plus is an internet magazine which aims to introduce readers to the beauty and the practical applications of mathematics.* The article concerned is at http://plus.maths.org/issue23/features/kissing/index.html and is an abridged chapter from George Szpiro's book *Kepler's Conjecture: How Some of the Greatest Minds in History Helped Solve One of the Oldest Math Problems in the World.* See http://www.amazon.co.uk/Keplers-Conjecture-Greatest-History-Problems/dp/product-description/0471086010.

Figure 111 – Icosahedron
From the English Wikipedia article *Icosahedron*, web page http://en.wikipedia.org/wiki/Icosahedron. GNU FDL image by DTR.

Figure 112 – Fibonacci spirals
Simple line drawing no image credit.

Figure 113 – The standard model
From the English Wikipedia article *Standard Model*, see web page http://en.wikipedia.org/wiki/Standard_Model. Image by TriTertButoxy.

Figure 114 – Torus knots
KnotPlot image courtesy of Doctor Robert Scharein. Again this is a crop. See the complete image at http://www.knotplot.com/knot-theory/torus_xing_.html.

Figure 115 – Standard model interactions
Again courtesy of *The Contemporary Physics Education Project*, see http://www.cpepweb.org/ where you can find out more. Also see *The Particle Adventure, a constantly evolving educational project sponsored by the Particle Data Group at Lawrence Berkeley National Laboratory (LBNL).*

Figure 116 – Trefoil warning symbol
From the English Wikipedia article *Trefoil*, see http://en.wikipedia.org/wiki/Trefoil. Image by Cary Bass.

Figure 117 – Expansion of the universe
This illustration of the "cosmic tug of war" between dark energy and dark matter is from the HubbleSite, see http://hubblesite.org/newscenter/archive/releases/nebula/2007/16/background/. Image credit: NASA/ESA/Ann Feild/STScl. Also see http://www.astronomy.com/asy/default.aspx?c=a&id=4675 for background reading.

Figure 118 – WMAP map of the universe
From http://map.gsfc.nasa.gov/, where you can read: *The Wilkinson Microwave Anisotropy Probe (WMAP) is a NASA Explorer mission. It has produced a wealth of precise and accurate cosmological information. WMAP produced the first full-sky map of the microwave sky with a resolution of under a degree, about the angular size of the moon.* Image credit: NASA/WMAP science team.

Figure 119 – Galaxies
This is the famous Hubble Ultra Deep Field image, from http://hubblesite.org/. See web page http://hubblesite.org/newscenter/archive/releases/2004/07/image/a/. Image credit: NASA, ESA, S. Beckwith (STScI) and the HUDF Team.

Figure 120 – Flat galactic rotation curve
From the Ohio State University Department of Astronomy website, and in particular from Professor Richard W Pogge's Astronomy 162 course notes, see http://www.astronomy.ohio-state.edu/~pogge/Ast162/Unit6/dark.html. Note that this is an illustration rather than actual data.

Figure 121 – Inverse square rule
Simple line drawing no image credit.

Figure 122 – Bottle-glass space
From the HubbleSite news centre, see web page http://hubblesite.org/newscenter/archive/releases/2007/17/ where the story starts: *Astronomers using NASA's Hubble Space Telescope have discovered a ghostly ring of dark matter that formed long ago during a titanic collision between two massive galaxy clusters.* Image credit: NASA, ESA, M.J. Jee and H. Ford (Johns Hopkins University).

Figure 123 – Dark matter pie
The upper portion is from an article by Bob Silberg entitled *SIM PlanetQuest to Predict Date of Cosmic Collision.* See http://www.nasa.gov/vision/universe/stars galaxies/Collision_Feature.html where you can read how *a team led by the University of Maryland's Ed Shaya will establish for the first time the masses and orbits of galaxies ranging from one million to 15 million light years from Earth.* SIM stands for Space Interferometry Mission, see http://planetquest.jpl.nasa.gov/SIM/sim_what_is.cfm. I edited the upper portion to make the lower portion.

Figure 124 – Stress ball
From English Wikipedia article *Stress ball*, see http://en.wikipedia.org/wiki/Stress_ball. Image by Kallemax.

Figure 125 – Giant black hole rips apart star
From the Chandra X-ray Observatory. See http://chandra.harvard.edu/ and the press release of 18th February 2004 at http://chandra.harvard.edu/press/04_releases/press_021804.html. The image is an artist's concept, and is credited to NASA, the

CXC, and Melissa Weiss again. As an aside, note that the mirrors on an X-ray telescope like Chandra are very different to those on an optical telescope:

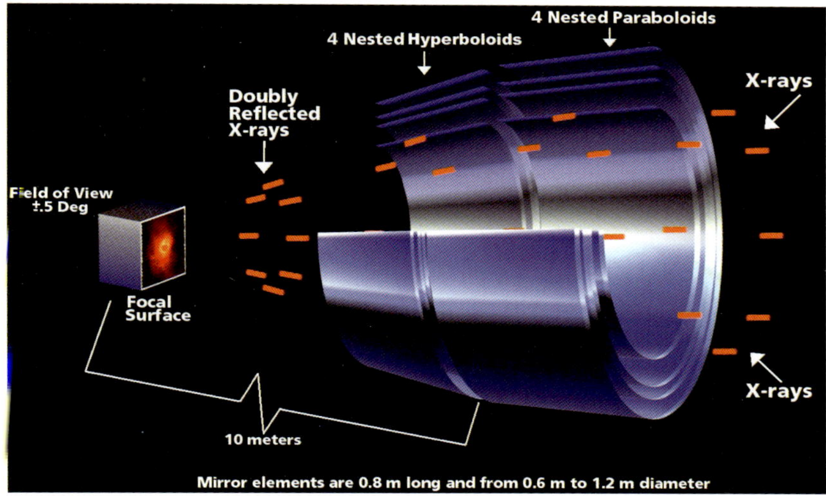

Figure 126 – Galilean relativity
Simple line drawing no image credit.

Figure 127 – Philosophical Transactions
This is *Philosophical Transactions 74* from 1784, containing John Michell's dark star paper. The image is courtesy of The Manhattan Rare Book Company. See http://www.manhattanrarebooks-science.com/images/black%20holes.jpg. I kid ye not, you still see references to this paper, which read: *Michell, J., 1784, "On the Means of discovering the Distance, Magnitude, etc. of the Fixed Stars, in consequence of the Diminution of the velocity of their Light, in case such a Diminution should be found to take place in any of them, and such Data should be procurred from Observations, as would be farther necessary for that Purpose", Philosophical Transactions, 74: 35-57.* Of course, whenever somebody mentions John Michell's "dark star", I can't help thinking of the cult John Carpenter movie featuring the kept-at-arms-length dysfunctional hippy crew and the pheno-menological "let there be light" bomb number 20. I first watched Dark Star in the Phoenix, East Finchley. It was a midnight movie. I loved it then, and I love it now. See http://en.wikipedia.org/wiki/ Dark_Star_(film).

Figure 128 – Black hole with accretion disk and jets
Credit: NASA. This image is from the Goddard Space Flight Center, see the article *Einstein makes extra dimensions toe the line* at http://www.nasa.gov/centers/goddard/news/topstory/2003/1212einstein.html. In particular see image 3 where the caption reads: *this artist's concept reveals the anatomy of a blazar. A blazar refers to*

a type of quasar, a bright galaxy with an active supermassive black hole at its core. Like quasars, blazars have particle jets, only the jets - by chance alignment - are pointed directly in Earth's direction.

Figure 129 – Triple alpha process
Image by Borb, see http://commons.wikimedia.org/wiki/Image:Triple-Alpha_Process.svg, reproduced under the terms of the GNU FDL.

Figure 130 – Main sequence
From an ESO press release *Back on Track* dated 19 June 2007, see http://www.eso.org/public/outreach/press-rel/pr-2007/pr-28-07.html. Also see the "about" page where you can read: *ESO is the European Organisation for Astronomical Research in the Southern Hemisphere. Created in 1962, ESO provides state-of-the-art research facilities to European astronomers and astrophysicists and is supported by Belgium, the Czech Republic, Denmark, Finland, France, Germany, Italy, the Netherlands, Portugal, Spain, Sweden, Switzerland, and the United Kingdom.*

Figure 131 – Neutron star with jet
From *Gamma-ray Bursts Active Longer Than Thought*, a press release on the Swift XRT and GRB 060714. It's from the NASA/Goddard Space Flight Center, and featured in ScienceDaily in May 2007. See http://www.sciencedaily.com/releases/2007/05/070523103847.htm for more. The caption reads: *This artwork depicts the central engine of a gamma-ray burst. A powerful jet of radiation and fast-moving particles blasts its way out of the central region of a dying star. The jet is presumably powered by material spiralling into a black hole or neutron star. Multiple episodes of infall provides fuel for the engine, leading to the burst and later X-ray flares. (Credit: NASA / SkyWorks Digital).*

Figure 132 – Infinite time dilation
From Thomas Knierim's *The Big View* website http://www.thebigview.com/, see page http://www.thebigview.com/spacetime/timedilation.html.

Figure 133 – Helical spring
From the English Wikipedia article http://en.wikipedia.org/wiki/Coil_spring. Image by Dvortygirl, reproduced under the terms of the GNU FDL.

Figure 134 – Simulated image of a stellar black hole
Image courtesy of Alain Riazuelo, a cosmologist at the Paris Astrophysics Institute. It used to feature in the English Wikipedia black hole article, see http://en.wikipedia.org/wiki/Black_hole, but at the time of writing is not shown. It is however on the French page at http://fr.wikipedia.org/wiki/Trou_noir.

Figure 135 – Hyperbolic
This is *Hyperbolic Order-3 heptakis heptagonal tiling* by Claudio Rocchini, reproduced (and colour-inverted) under the terms of the GNU FDL. See English Wikipedia article http://en.wikipedia.org/wiki/Uniform_tilings_in_hyperbolic_plane.

Figure 136 – Time line of the Universe
Image credit: NASA/WMAP science team. I saw this picture in the article *Ringside seat to the universe's first split second* at http://www.nasa.gov/vision/universe/stars galaxies/wmap_pol.html. Also see http://map.gsfc.nasa.gov/news/ where the caption reads: *A representation of the evolution of the universe over 13.7 billion years. The far left depicts the earliest moment we can now probe, when a period of "inflation" produced a burst of exponential growth in the universe. (Size is depicted by the vertical extent of the grid in this graphic.) For the next several billion years, the expansion of the universe gradually slowed down as the matter in the universe pulled on itself via gravity. More recently, the expansion has begun to speed up again as the repulsive effects of dark energy have come to dominate the expansion of the universe. The afterglow light seen by WMAP was emitted about 380,000 years after inflation and has traversed the universe largely unimpeded since then.*

Figure 137 – Hawking radiation warning sign
This is a *laser* warning sign, not a Hawking radiation warning sign. Sorry.

Figure 138 – Hawking radiation
Simple line drawing no image credit. But do a search on "powerful enough to rip apart the fabric of space and time", and see the *Extreme Light Infrastructure Project* on http://www.extreme-light-infrastructure.eu/High-field_5_2.php.

Figure 139 – Alfred Nobel
Copyright © The Nobel Foundation, see http://nobelprize.org/alfred_nobel/. This is the best place to read about Alfred Nobel, who was quite a guy: *Nobel received his first patent at age 30. Ten years later, he would establish factories in several countries and was the first person in the world to create an international holding company. The inventor of dynamite held 355 patents in several countries. As an entrepreneur, Nobel was unbeatable in his time.*

Figure 140 – Helaman Ferguson sculptures
From Helaman Ferguson's website, see http://www.helasculpt.com/gallery/index .html. Beautiful stuff. For the record, image 1 is *Eightfold Way* and includes a circle-limit type base. Image 2 is *Umbilical Torus*, featuring a radial-cross-section hypo-cycloid with three cusps, a planar-cross-section cardioid, and a space-filling hilbert curve articulated surface. Image 3 is *Figureight Knot Complement*, the reference being to the figure eight knot and its hyperbolic space complement. Image 4 is *Umbilic Torus (1,3) Yin.* It isn't the trefoil it might look like at first glance, but is actually an inverse of the umbilical torus, featuring a radial-cross-section cardioid and a planar-cross-section hypocycloid with three cusps.

Figure 141 – Calabi-Yau manifold
From English Wikipedia article *String theory*, see http://en.wikipedia.org/wiki/ String_theory. Image by Lunch. Also see http://en.wikipedia.org/wiki/Calabi-Yau_manifold and http://en.wikipedia.org/wiki/Superstring_theory.

Figure 142 – Tapestry of the Centuries
Image copyright © 1999 Vladimir Gorsky / Korab Inc. It's a world-famous painting, see http://www.gorskyfineart.com/ where you can find out more: *The original canvas measuring 9' x 18' took 3 years to paint and comprises 350 people and events that shaped the history of the world.*

Figure 143 – Rocket power
From Gary Novak's website http://nov55.com/, see page http://nov55.com/graf.html Gary offers some thought-provoking stuff, like: *The result is that the laboratory scientists are dominated by office scientists who dictate how their work will be designed and reported.*

Figure 144 – About Time, The End of Time, A World without Time
Jacket images courtesy of the publishers. *About Time* by Paul Davies was published by Viking, an imprint of Penguin Press in 1995. *The End of Time* by Julian Barbour was published by Weidenfeld and Nicolson, an imprint of the Orion Publishing Group in 1999. *A World without Time* by Palle Yourgrau was published by Basic Books in 2005.

Figure 145 – Energy
This image was created by MIT faculty members Jefferson Tester, Elisabeth Drake, Frank Incropera, and Michael Golay. It's from *course materials for 10.391J /1.818J /2.65J/11.371J/ 22.811J/ESD.166J, Sustainable Energy, IAP 2007 to Spring 2007, © 2007 MIT OpenCourseWare (http://ocw.mit.edu), Massachusetts Institute of Technology.* Downloaded on 31 August 2008, reproduced under the terms and conditions of the OCW Creative Commons license, see http://ocw.mit.edu/Ocw Web/web/terms/terms/index.htm. Also see web page http://ocw.mit.edu/OcwWeb/ Chemical-Engineering/10-391JJanuary--IAP--2007-Spring-2007/CourseHome/index .htm for the image itself. OCW is free lecture notes, exams, and videos, and is absolutely commendable.

Figure 146 – Angular mass
From http://motivate.maths.org/, see Dr Hugh Hunt's *The Wonderful World of Gyroscopes* on web page http://motivate.maths.org/conferences/conf14/c14_talk4 .shtml. "Motivate" is another commendable initiative, *providing maths, science and cross-curricular videoconferences and linked projects for students of all ages (5-19) both in the UK and internationally.* Also see Hugh's personal webpage at http://www2.eng.cam.ac.uk/~hemh/. Hugh is a senior lecturer in Mechanics within the Department of Engineering at the University of Cambridge, and a Fellow and College Lecturer in Engineering at Trinity College.

Figure 147 – Molecular electrostatic potential
From *Chemistry Software*, see http://www.chemistry-software.com/modelling /10190.htm. The ChemSite 3D molecular visualization software lets you *create stunning presentations with photo-realistic rendering tools such as Bezier curves, Ray-tracing, Phong shading and texture-mapped atom symbols.*

Figure 148 – Gravity mesh
I bumped into this image on Fraser Cain's *Universe Today* website. Check out his Gravity Probe B article at http://www.universetoday.com/2004/04/23/gravity-probe-b-is-working-fine/. Gravity probe B is a NASA mission, see http://www.nasa.gov/mission_pages/gpb/index.html for further details. The image itself is credited to NASA and is entitled *Gravity Probe B and Space-Time.*

Figure 149 – Silly putty in a vacuum chamber
From *Crazy Aaron's Puttyworld.* See http://www.puttyworld.com/ and the fun pictures page at http://www.puttyworld.com/funpictures.html. The image is *Thinking Putty in a Vacuum Chamber: Matthew Cole from Kent, UK was bored at work and put his Thinking Putty in a vacuum chamber (don't we all have one at work!) After exposing it to the emptiness of outer space, he was kind enough to send us a photo. There's more. Like Strange Attractor shown below. Wow:*

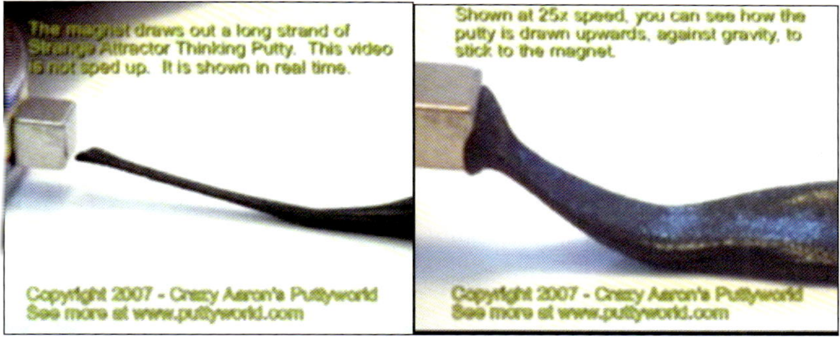

Figure 150 – STAR particle collision
From http://www.bnl.gov/rhic/STAR.htm. STAR is the Soleniodal Tracker detector At the Relativistic heavy ion collider at the Brookhaven National Laboratory. See http://www.bnl.gov/RHIC/full_en_images.htm. The captions reads: *one of the first full-energy collisions between gold ions at Brookhaven Lab's Relativistic Heavy Ion Collider, as captured by the Solenoidal Tracker At RHIC (STAR) detector. The tracks indicate the paths taken by thousands of subatomic particles produced in the collisions as they pass through the STAR Time Projection Chamber, a large, 3-D digital camera.*

Figure 151 – The universe (SDSS)
I first saw this image in an article entitled *Lighting up the Dark Universe* on the Los Alamos web site, see http://www.lanl.gov/science/1663/universe.php. The caption reads: *A Sloan Digital Sky Survey map of the luminous universe out to a distance of 5 billion light years. We (and the present time) are located at the center between the two pie-shaped regions. Photo courtesy of the Sloan Digital Sky Survey.* See the Sloan Digital Sky Survey website *http://www.sdss.org/* where you can read: *In 2006, SDSS advanced mankind's understanding of the universe with several new*

discoveries. The survey found new dwarf companion galaxies to the Milky Way; confirmed Einstein's prediction of cosmic magnification; observed the largest known structures in the universe (measuring more than a billion light years across); and further unraveled our galaxy's active past, filled with galactic mergers. (Funding for the SDSS and SDSS-II has been provided by the Alfred P. Sloan Foundation, the Participating Institutions, the National Science Foundation, the U.S. Department of Energy, the National Aeronautics and Space Administration, the Japanese Monbukagakusho, the Max Planck Society, and the Higher Education Funding Council for England. The SDSS is managed by the Astrophysical Research Consortium for the Participating Institutions. The Participating Institutions are the American Museum of Natural History, Astrophysical Institute Potsdam, University of Basel, University of Cambridge, Case Western Reserve University, University of Chicago, Drexel University, Fermilab, the Institute for Advanced Study, the Japan Participation Group, Johns Hopkins University, the Joint Institute for Nuclear Astrophysics, the Kavli Institute for Particle Astrophysics and Cosmology, the Korean Scientist Group, the Chinese Academy of Sciences (LAMOST), Los Alamos National Laboratory, the Max-Planck-Institute for Astronomy (MPIA), the Max-Planck-Institute for Astrophysics (MPA), New Mexico State University, Ohio State University, University of Pittsburgh, University of Portsmouth, Princeton University, the United States Naval Observatory, and the University of Washington.)

Figure 152 – Black-hole microquasar XTE J1118+480
From a 2001 press release issued by the National Radio Astronomy Observatory station in Socorro, see http://www.nrao.edu/pr/2001/blackhole/highvbh.graphics.html. Image credit: I. Rodrigues and I.F. Mirabel, Space Telescope Science Institute, NRAO/AUI/NSF. The Very Large Array run from Socorro was featured in the movie *Contact* starring Jodie Foster. See http://www.aoc.nrao.edu/.

Figure 153 – Peer review process
Courtesy and copyright Nick Kim. See http://www.lab-initio.com/. Nick is an environmental chemist in New Zealand as well as a cartoonist.

Figure 154 – Gulliver
From the Project Gutenberg eBook *Young Folks Treasury, Volume 3 (of 12), by Various, edited by Hamilton Wright Mabie,* reproduced under the terms of the Project Gutenberg License: http://www.gutenberg.org/license. See the Project Gutenberg main page at http://www.gutenberg.net/ and follow the "about us" link to read the history: *Project Gutenberg began in 1971 when Michael Hart was given an operator's account with $100,000,000 of computer time in it by the operators of the Xerox Sigma V mainframe at the Materials Research Lab at the University of Illinois. This was totally serendipitous, as it turned out that two of a four operator crew happened to be the best friend of Michael's and the best friend of his brother.* Hats off to Michael Hart. And hats off to Project Gutenberg.

Ciao.